新工科建设之路·计算机类专业应用型特色教材

Android 应用开发案例教程

卢向华　郭玉珂　郑卫东　主　编

电子工业出版社

Publishing House of Electronics Industry

北京·BEIJING

内 容 简 介

本书采用的开发环境是 Android Studio 4.1 和 Android 11.0，以 Android 应用开发为主线，介绍基于 Android 移动应用软件开发的相关技术。内容涵盖了 Android Studio 开发环境搭建、UI 编程、Activity、Intent、数据存储、ContentProvider、Service、BroadcastReceiver、网络与数据处理、应用项目开发等。

本书注重理论与实践的结合，每章都提供了大量的实例，所有的知识点均有理论解析和实际应用，使读者能够快速理解并掌握相关知识在实际开发中的应用。第 10 章通过一个完整的实际项目，运用软件工程的设计思想，介绍 Android 项目的开发过程，使读者能够真正把本书的知识应用到实际开发中，全面提高分析问题、解决问题和编写代码的能力。

本书既可作为高等院校本科、专科计算机相关专业 Android 开发的教材，也可作为 Android 爱好者和开发人员的参考书。

未经许可，不得以任何方式复制或抄袭本书之部分或全部内容。
版权所有，侵权必究。

图书在版编目（CIP）数据

Android 应用开发案例教程/卢向华，郭玉珂，郑卫东主编. —北京：电子工业出版社，2021.9
ISBN 978-7-121-41975-1

Ⅰ.①A… Ⅱ.①卢… ②郭… ③郑… Ⅲ.①移动终端－应用程序－程序设计－高等学校－教材
Ⅳ.①TN929.53

中国版本图书馆 CIP 数据核字（2021）第 186414 号

责任编辑：戴晨辰　　　文字编辑：王　炜
印　　刷：北京雁林吉兆印刷有限公司
装　　订：北京雁林吉兆印刷有限公司
出版发行：电子工业出版社
　　　　　北京市海淀区万寿路 173 信箱　邮编：100036
开　　本：787×1 092　1/16　印张：18.25　字数：467.2 千字
版　　次：2021 年 9 月第 1 版
印　　次：2021 年 9 月第 1 次印刷
定　　价：59.00 元

凡所购买电子工业出版社图书有缺损问题，请向购买书店调换。若书店售缺，请与本社发行部联系，联系及邮购电话：（010）88254888，88258888。
质量投诉请发邮件至 zlts@phei.com.cn，盗版侵权举报请发邮件至 dbqq@phei.com.cn。
本书咨询联系方式：dcc@phei.com.cn。

随着 Android 生态的发展壮大，Android 系统已经逐步渗透到平板电脑、智能电视、车载大屏、智能家居等领域。目前 Android 系统的全球市场份额已接近 90%，在全球范围内占据主导地位。随着 5G 网络的持续铺开，未来 Android 的市场规模有望进一步拓展，尤其在物联网相关领域会爆发出大量的市场需求，必将带动对 Android 开发人才需求的再度高涨。基于 Android 的移动应用开发是当前移动互联网开发领域的一个重要方向，目前大多数高校计算机相关专业均开设有移动应用开发课程。

谷歌公司推出 Android 的新版本除增加一些新特性、新功能外，在管理机制、控制策略等方面也有根本性的变化。另外，移动互联网领域更是发展迅速，新技术、新应用不断出现，而现有教材内容更新迟缓，无法跟上 Android 系统的发展，也不能及时体现前沿技术应用和行业发展趋势。虽然市场上与 Android 开发相关的书籍比较多，但大多数是针对有一定基础的业内开发人员编写的，并不完全符合高校的教学要求。

编者根据多年移动应用软件课程教学、项目开发经验，调研多所高等院校 Android 应用开发课程设置情况，结合企业对 Android 应用开发人才专业技能的要求，增加了针对 Android 系统新功能的知识应用，并适当加入实际开发中广泛应用的技术和第三方开源库，为学生提供具有时效性、针对性的课堂知识。通过在课堂教学中适当引入学科发展前沿和产业发展现状，在实践性教学中融入技能大赛的相关成果，让学生了解最新的产业动向，紧跟产业潮流，找到与产业发展的连接点，为职业学习和创新创业寻找突破口。

本书特点如下。

（1）本书内容紧随技术发展，以能力培养为导向，适应工程教育人才培养课程的改革要求。采用 Android Studio 4.1 开发工具和谷歌公司发布的最新 SDK API30，摈弃所有过时的内容及相关技术，使学生所学的技术不会在短时间内过时。

（2）本书内容安排循序渐进、由易到难，符合初学者的认识规律和学习路径。同时尽可能地采取步骤教学法，非常适合从未接触过 Android 开发的学生学习。

（3）理论与实践相结合，有利于学生快速理解并掌握相关知识在实际开发中的应用。每章配套的案例设计经典，所有案例均提供完整的代码，并全部在 Android Studio 4.1 开发环境中调试通过，方便学生单独试做。

（4）教学资源丰富，配套有教学课件、所有实例及章末习题的完整源代码、微视频，方便教师教学使用。登录华信教育资源网（www.hxedu.com.cn），注册后可免费下载本书相关配套资源。

本书由洛阳理工学院的卢向华、郭玉珂、郑卫东编写。全书共 10 章，第 1 章、第 4 章、第 6 章、第 7 章和第 8 章由卢向华编写，第 2 章、第 3 章和第 9 章由郭玉珂编写，第 5 章和第 10 章由郑卫东编写，全书由卢向华统稿。

由于编者水平有限及技术的快速迭代，书中的疏漏和不足在所难免，敬请专家与读者批评指正。

编 者

第 1 章 Android 入门 ······1
1.1 走进 Android ······1
1.1.1 Android 发展史 ······1
1.1.2 Android 系统架构 ······2
1.2 Android 开发环境搭建 ······4
1.2.1 Android 开发环境的配置要求 ······4
1.2.2 JDK 的下载安装与环境变量的配置 ······4
1.2.3 Android Studio 的下载安装 ······6
1.3 开发 Android 应用程序 ······14
1.3.1 创建 Android 应用程序 ······14
1.3.2 Android 应用程序目录结构 ······17
1.3.3 创建 Android 模拟器 ······17
1.3.4 Android 程序的运行和打包 ······20
习题 1 ······23

第 2 章 用户界面设计基础 ······24
2.1 用户界面编写方式 ······24
2.2 常用布局 ······26
2.2.1 布局通用属性 ······26
2.2.2 LinearLayout ······26
2.2.3 RelativeLayout ······28
2.2.4 FrameLayout ······30
2.2.5 TableLayout ······31
2.2.6 GridLayout ······33
2.2.7 ConstraintLayout ······34
2.2.8 AbsoluteLayout ······37
2.3 常用控件 ······37

2.3.1 TextView 控件 ······38
2.3.2 EditText 控件 ······39
2.3.3 Button 控件 ······41
2.3.4 ImageView 控件 ······43
2.3.5 RadioButton 控件 ······44
2.3.6 CheckBox 控件 ······46
2.3.7 Toast 控件 ······48
习题 2 ······51

第 3 章 用户界面高级控件 ······53
3.1 弹出式控件 ······53
3.1.1 AlertDialog 控件 ······53
3.1.2 Notification 控件 ······60
3.2 日期/时间选择器 ······62
3.2.1 DatePicker 控件 ······63
3.2.2 TimePicker 控件 ······64
3.3 滚动条和进度条 ······65
3.3.1 ScrollView 控件 ······66
3.3.2 ProgressBar 控件 ······69
3.4 列表视图 ······69
3.4.1 ListView 控件 ······70
3.4.2 适配器 ······72
3.5 自定义控件 ······74
习题 3 ······76

第 4 章 程序基本单元 Activity ······78
4.1 Activity 概述 ······78
4.2 Activity 的生命周期 ······78
4.2.1 生命周期状态 ······78

4.2.2 生命周期方法 79
4.3 Activity 的使用 83
　　4.3.1 创建 Activity 83
　　4.3.2 配置 Activity 85
　　4.3.3 启动 Activity 和关闭 Activity 86
4.4 Intent 与 IntentFilter 89
　　4.4.1 Intent 89
　　4.4.2 IntentFilter 94
4.5 多个 Activity 的使用 97
　　4.5.1 Activity 之间数据的传递 98
　　4.5.2 Activity 之间数据的回传 103
4.6 使用 Fragment 111
　　4.6.1 Fragment 的生命周期 111
　　4.6.2 创建 Fragment 113
　　4.6.3 在 Activity 中添加 Fragment 114
　　4.6.4 Activity 与 Fragment 的通信 118
习题 4 123

第 5 章 Android 数据存储 125

5.1 SharedPreferences 数据存储 125
　　5.1.1 使用 SharedPreferences 存储数据 125
　　5.1.2 使用 SharedPreferences 读取数据 127
　　5.1.3 SharedPreferences 使用示例 127
　　5.1.4 SharedPreferences 使用注意事项 128
5.2 Android 权限管理 129
　　5.2.1 权限机制 129
　　5.2.2 运行时权限申请 130
5.3 数据的文件存储 133
　　5.3.1 Android 文件存储概述 133
　　5.3.2 文件的内部存储 134
　　5.3.3 文件的外部存储 135
　　5.3.4 文件存储操作示例 137
5.4 数据库 SQLite 148
　　5.4.1 SQLite 数据库简介 149
　　5.4.2 创建 SQLite 数据库 149
　　5.4.3 数据库操作的实现 150
　　5.4.4 SQLite 数据库使用示例 152
习题 5 159

第 6 章 内容提供者 160

6.1 ContentProvider 简介 160
6.2 URI 简介 160
6.3 开发 ContentProvider 162
　　6.3.1 创建和注册 ContentProvider 162
　　6.3.2 使用 ContentResolver 操作数据 164
6.4 监听 ContentProvider 的数据改变 169
6.5 使用系统内置的 ContentProvider 172
习题 6 175

第 7 章 服务 176

7.1 Service 简介 176
7.2 Service 的生命周期 176
7.3 Service 的使用 179
　　7.3.1 创建和配置 Service 179
　　7.3.2 使用 startService() 方法启动 Service 180
　　7.3.3 使用 bindService() 方法启动 Service 185
　　7.3.4 Service 与 Activity 的通信 186
7.4 访问系统服务 192
7.5 异步消息处理 197
　　7.5.1 Handler 消息传递机制 197
　　7.5.2 AsyncTask 类 203
习题 7 206

第 8 章 广播机制 207

8.1 Android 系统的广播机制 207
8.2 BroadcastReceiver 208
　　8.2.1 广播接收器的创建 208

		8.2.2 广播接收器的注册 ……………209

8.3　接收系统广播 ……………………211
8.4　自定义广播 ………………………213
　　8.4.1　广播类型 …………………213
　　8.4.2　普通广播 …………………214
　　8.4.3　有序广播 …………………215
8.5　本地广播 …………………………217
习题 8 ……………………………………218

第 9 章　网络编程 …………………………219

9.1　使用 HTTP 访问网络 ……………219
　　9.1.1　网络编程基本概念 …………219
　　9.1.2　使用 HttpURLConnection
　　　　　连接网络 ……………………220
　　9.1.3　网络信息传输 ………………223
　　9.1.4　XML 和 JSON ………………227
9.2　Android 网络访问框架 …………231
　　9.2.1　Volley ………………………231
　　9.2.2　OkHttp ……………………234
　　9.2.3　WebView …………………236
9.3　Socket 网络编程 …………………240
习题 9 ……………………………………247

第 10 章　社区服务系统 ……………………248

10.1　项目简介 …………………………248
10.2　功能需求 …………………………248
　　10.2.1　Android 手机端 ……………248
　　10.2.2　Web 服务器端 ……………250
10.3　效果展示 …………………………250
10.4　系统设计与实现 …………………252
　　10.4.1　数据库设计 ………………252
　　10.4.2　Web 服务器端设计 ………253
　　10.4.3　Android 手机端的设计
　　　　　　与实现 ………………………259

第 1 章 Android 入门

Android 是由谷歌公司和开放手机联盟共同开发的一种基于 Linux 的移动设备操作系统，它自问世以来备受关注，已成为移动平台最受欢迎的操作系统之一。随着物联网、移动互联网和人工智能技术的蓬勃发展，Android 系统除在智能手机领域外，还在智能硬件、智能物联、穿戴设备等领域得到了大规模推广。本章将介绍 Android 发展史、Android 系统架构、Android 开发环境搭建，以及 Android 应用程序的创建、调试和发布等内容。

1.1 走进 Android

1.1.1 Android 发展史

Android 最初由 Andy Rubin 开发，于 2005 年 8 月被谷歌公司收购。2007 年 11 月 5 日，谷歌公司正式向外界展示了这款操作系统，并且宣布建立一个全球性开放手机联盟（Open Handset Alliance）来共同研发改良 Android 系统。这个联盟将支持谷歌公司发布的手机操作系统及应用软件。谷歌公司以 Apache 免费开源许可证的授权方式，发布了 Android 的源代码。

2008 年，在 Google I/O 大会上，谷歌公司提出了 Android HAL 的架构方案，同年 8 月，Android 获得了美国联邦通信委员会（FCC）的批准。一个月后，谷歌公司正式发布了 Android 1.0 系统，这也是 Android 系统最早的版本。

2009 年 4 月，谷歌公司正式推出 Android 1.5 后，Android 版本开始以甜品的名字命名，Android 1.5 命名为 Cupcake（纸杯蛋糕）。

2009 年 9 月，谷歌公司发布了 Android 1.6 Donut（甜甜圈）的正式版，并且推出了搭载 Android 1.6 的手机 HTC Hero（G3）。凭借着出色的外观设计及全新的 Android 1.6，HTC Hero（G3）成为当时全球最受欢迎的手机之一。

2009 年 10 月，谷歌公司发布 Android 2.0/2.1 Eclair（松饼）。

2010 年 5 月，谷歌公司正式发布 Android 2.2 Froyo（冻酸奶）。

2010 年 10 月，谷歌公司宣布 Android 系统达到了第一个里程碑，即电子市场上获得官方数字认证的 Android 应用数量已经达到 10 万个，Android 系统的应用增长非常迅速。

2010 年 12 月，谷歌公司正式发布 Android 2.3 Gingerbread （姜饼）。

2011 年 2 月，谷歌公司发布 Android 3.0 Honeycomb（蜂巢），这是第一个 Android 平板操作系统，并于同年 5 月和 7 月分别发布了 Android 3.1 和 Android 3.2。

2011 年 1 月，谷歌公司称 Android 设备每日的新增销售量达到 30 万部，2011 年 7 月，这个数字增长到 55 万部，而 Android 系统设备的用户总数已达到 1.35 亿人，Android 系统成为智能手机领域占有量最高的系统。2011 年 8 月，Android 手机已占据全球智能手机市场 48%的份额，并在亚太地区市场占据统治地位，终结了 Symbian（塞班系统）的霸主地位，跃居全球第一。

2011 年 10 月，谷歌公司发布全新的 Android 4.0 Ice Cream Sandwich（冰激凌三明治）。Android 3.0 之前的版本主要针对移动手机，Android 蜂巢版本系列（3.0、3.1 和 3.2 的版本）主要针对平板电脑及上网本，而 Android 4.0 之后的版本将同时支持移动手机、平板电脑及上网本等终端。

2012 年 6 月，谷歌公司发布 Android 4.1 Jelly Bean（果冻豆），同年 10 月推出 Android 4.2 Jelly Bean（果冻豆）。

2013 年 11 月，谷歌公司正式发布 Android 4.4 KitKat（奇巧巧克力）。

2014 年 10 月，谷歌公司发布 Android 5.0 Lollipop（棒棒糖）。

2015 年 9 月，谷歌公司发布 Android 6.0 Marshmallow（棉花糖），Android 6.0 在对软件体验与运行性能上进行了大幅度的优化。

2016 年 8 月，谷歌公司发布 Android 7.0 Nougat（牛轧糖）。

2017 年 8 月，谷歌公司发布 Android 8.0 Oreo（奥利奥）。

2018 年 5 月，谷歌公司发布 Android 9.0 Pie（派）。Android 9.0 充满了人工智能元素，包括更改通知栏样式，以及在整体设计中添加了更多的圆形元素。

2019 年 8 月，谷歌公司宣布 Android 系统的重大改变，更换了全新的 Logo，也不再以甜品命名。同年 9 月，谷歌公司正式发布的 Android 10.0 命名为 Android Q。

2020 年 9 月，谷歌公司发布 Android 11.0，按照 26 个英文字符的自然顺序演进，将 Android 11.0 命名为 Android R。

1.1.2　Android 系统架构

Android 系统采用分层架构，由低到高分为 4 层，分别是 Linux 内核、核心类库、应用框架层和应用层，如图 1.1 所示。

1. Linux 内核（LINUX KERNEL）

Android 的核心系统服务是基于 Linux 内核的，如安全性、内存管理、进程管理、网络协议栈和驱动模型等都依赖于该内核。Linux 内核同时也作为硬件和软件栈之间的抽象层，为 Android 设备的各种硬件提供了底层驱动，其主要驱动如下。

- Display Driver：显示驱动，基于 Linux 的帧缓冲驱动。
- Camera Driver：照相机驱动，基于 Linux 的 v412 驱动。
- Bluetooth Driver：蓝牙驱动，基于 IEEE 802.15.1 标准的无线传输技术。
- Flash Memory Driver：Flash 闪存驱动，基于 MTD 的 Flash 驱动程序。
- Binder（IPC）Driver：Android 的一个特殊驱动程序，具有单独的设备节点，可提供进程间通信的功能。
- USB Driver：USB 接口驱动。

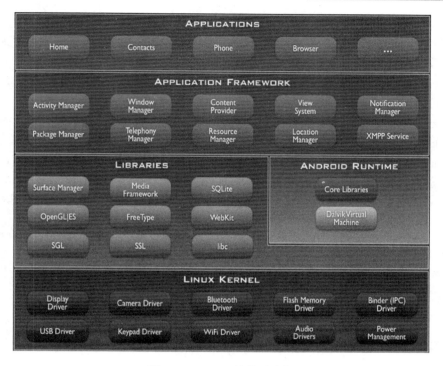

图 1.1　Android 的体系结构

- Keypad Driver：键盘驱动，可作为输入设备的键盘驱动。
- WiFi Driver：基于 IEEE 802.11 标准的驱动程序。
- Audio Drivers：音频驱动，基于 ALSA（Advanced Linux Sound Architecture）的高级 Linux 声音体系驱动。
- Power Management：电源管理，如电池、电量等。

2．核心类库（LIBRARIES）

核心类库包含系统库及 Android 运行环境。

系统库通过 C/C++库为 Android 系统提供主要的特性支持，如 SQLite 库提供数据库的支持，OpenGL|ES 库提供 3D 绘图的支持，Webkit 库提供浏览器内核的支持等。

Android 运行时，一些核心库能够允许开发者使用 Java 语言来编写 Android 应用。另外，Android 运行时，Dalvik 虚拟机能使每个 Android 应用都运行在独立的进程中，并且拥有一个自己的 Dalvik 虚拟机实例。Dalvik 虚拟机是专门为移动设备定制的，并针对手机内存、CPU 性能有限等情况进行了优化处理。

3．应用框架层（APPLICATION FRAMEWORK）

应用框架层主要提供构建应用程序时用到的各种 API，开发人员只要使用这些框架来开发自己的应用程序，就可以简化程序开发的架构设计。Android 应用框架层提供的主要 API 框架如下。

Activity Manager：活动管理器，用来管理应用程序生命周期，并提供常用的导航退回功能。

Window Manager：窗口管理器，用来管理所有的窗口程序。

Content Provider：内容提供者，它可以让一个应用访问另一个应用的数据，或者共享它们自己的数据。

View System：视图管理器，用来构建应用程序，如列表、表格、文本框及按钮等。

Notification Manager：通知管理器，用来设置在状态栏中显示的提示信息。

Package Manager：包管理器，用来对 Android 系统内的程序进行管理。

Telephony Manager：电话管理器，用来对联系人及通话记录等信息进行管理。

Resource Manager：资源管理器，用来提供非代码资源的访问，如本地字符串、图形及布局文件等。

Location Manager：位置管理器，用来提供使用者的当前位置等信息，如 GPRS 定位。

XMPP Service：XMPP 服务。

4．应用层（APPLICATIONS）

应用层是用 Java 语言编写的、运行在 Android 平台上的程序，如 SMS 短信、日历、地图及浏览器等程序。Android 开发人员就是要编写在应用层上运行的应用程序。

1.2 Android 开发环境搭建

1.2.1 Android 开发环境的配置要求

Android Studio 是谷歌公司开发的一款面向 Android 开发者的集成开发环境，支持 Windows、Mac、Linux 等操作系统。

目前，Android 开发的主流语言是 Java，因此 Android 开发需要安装 JDK 和 Android SDK （Software Development Kit）。Android SDK 是谷歌公司提供的 Android 开发工具包。

Android 各个版本的 SDK 包都较大，且运行 Android 模拟器也要耗费大量内存，因此安装 Android Studio 开发环境对 CPU、内存和硬盘的要求较高，建议内存至少 8GB 及以上，硬盘（尤其是系统盘）最好是大容量固态硬盘。

Android Studio 要求 Java JDK 1.7 为最低版本，读者可以通过 Oracle 官网下载适合自己系统的最新版本。

1.2.2 JDK 的下载安装与环境变量的配置

1．下载和安装

在 Oracle 官网中下载适合自己系统的 JDK 安装包，下载完成后双击安装包，启动安装向导进行安装。

2．配置环境变量

下面以 Windows 7 系统为例配置环境变量。

（1）在桌面上，右击"计算机"图标，并在弹出的菜单中选择"属性"选项，打开系统窗口，选择"高级系统设置"选项，打开"系统属性"对话框，如图 1.2 所示。

（2）单击"环境变量"按钮，打开如图 1.3 所示的"环境变量"对话框。

图 1.2 "系统属性"对话框　　　　　　图 1.3 "环境变量"对话框

（3）单击"系统变量"选区下方的"新建"按钮，打开如图 1.4 所示的"新建系统变量"对话框。

图 1.4 "新建系统变量"对话框

（4）在"变量名"编辑框中输入"JAVA_HOME"，在"变量值"编辑框中输入 JDK 的安装路径，然后单击"确定"按钮。笔者将 JDK 安装在"D:\Program Files\Java\jdk-11.0.2"目录下，环境变量 JAVA_HOME 的设置如图 1.5 所示。

图 1.5 新建 JAVA_HOME 变量

（5）在"环境变量"对话框的"系统变量"选区中选择"Path 变量"选项，如图 1.6 所示，然后单击"编辑"按钮，打开"编辑系统变量"对话框，如图 1.7 所示。

（6）在"变量值"编辑框的最后部分添加";%JAVA_HOME%\bin"，然后单击"确定"按钮，完成环境变量的配置。

图 1.6 "环境变量"对话框　　　　图 1.7 "编辑系统变量"对话框

1.2.3　Android Studio 的下载安装

1．下载 Android Studio

可以在 Android Studio 中文社区官网下载适合自己系统版本的 Android Studio 安装包，如图 1.8 所示。

图 1.8　Android Studio 中文社区

2．安装 Android Studio 和 SDK

Android Studio 安装包下载完成后，按如下步骤进行安装。

（1）双击安装文件启动安装向导，其欢迎安装界面如图 1.9 所示。

（2）单击"Next"按钮，进入"Choose Components"界面，如图 1.10 所示。

（3）采用默认安装方式，单击"Next"按钮，进入"Configuration Settings"界面，如图 1.11 所示。

（4）单击"Browse"按钮，选择 Android Studio 的安装路径，然后单击"Next"按钮，进入"Choose Start Menu Folder"界面，如图 1.12 所示。

图 1.9　Android Studio 的欢迎安装界面

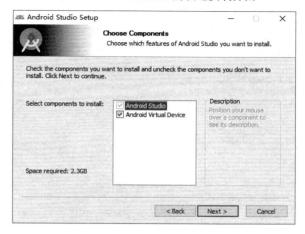

图 1.10　选择安装组件

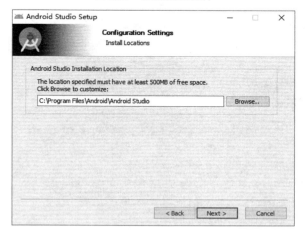

图 1.11　配置安装路径

（5）采用默认配置，将 Android Studio 的快捷方式创建在开始菜单中的指定文件夹下。单击"Install"按钮，开始安装，安装过程如图 1.13 所示。

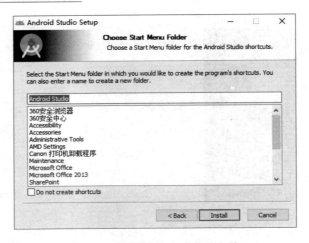

图 1.12　配置快捷方式菜单文件

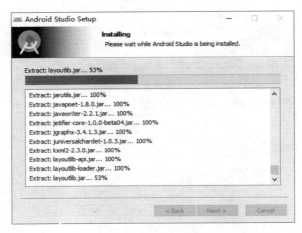

图 1.13　安装过程

（6）安装完成后，"Next"按钮变为可操作状态，如图 1.14 所示。单击"Next"按钮，打开如图 1.15 所示的界面，Android Studio 安装成功。

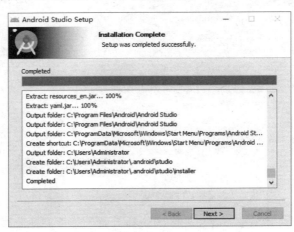

图 1.14　安装完成

（7）单击"Finish"按钮，此处默认勾选"Start Android Studio"复选框，启动 Android Studio，弹出如图 1.16 所示的"Import Android Studio Settings From…"对话框。

图 1.15　安装成功

图 1.16　导入配置

（8）选中"Do not import settings"单选按钮，单击"OK"按钮，进入 Android Studio 启动加载界面，如图 1.17 所示。

需要注意的是，在第一次启动时，Android Studio 将自动在线下载 SDK 并进行安装。此过程需要较长时间，应保持网络状态的稳定。

（9）加载完成后弹出如图 1.18 所示的"Android Studio First Run"对话框，询问是否设置代理，单击"Cancel"按钮，进入如图 1.19 所示的安装向导欢迎界面。

图 1.17　启动加载

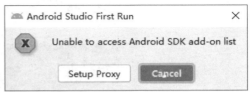

图 1.18　设置代理

（10）在图 1.19 所示的界面中，单击"Next"按钮，进入如图 1.20 所示的"Install Type"选择安装方式界面。

（11）选中"Standard"单选按钮，单击"Next"按钮，进入"Select UI Theme"主题设置界面，如图 1.21 所示。

图 1.19 安装向导欢迎界面

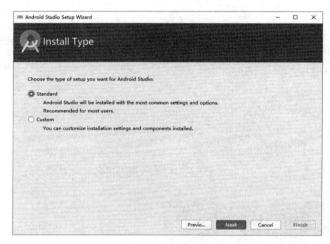

图 1.20 选择安装方式界面

图 1.21 主题设置界面

（12）选择自己喜欢的主题风格，单击"Next"按钮，进入"SDK Components Setup"界面，如图 1.22 所示。

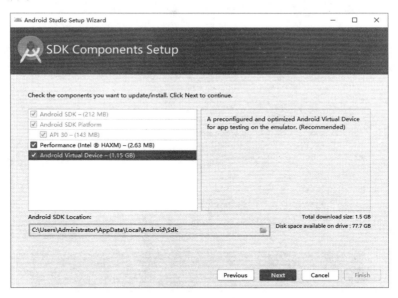

图 1.22　SDK 组件选择

（13）设置下载的 SDK 组件和保存路径，单击"Next"按钮进入"Verify Settings"界面，确认 SDK 信息，如图 1.23 所示。

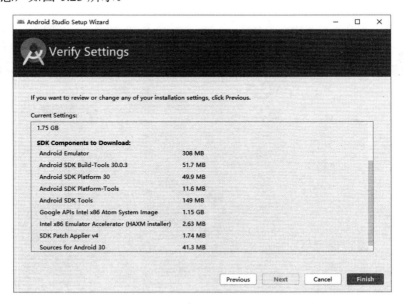

图 1.23　SDK 信息

（14）单击"Finish"按钮，开始下载 SDK，如图 1.24 所示，此过程需要的时间较长。

（15）SDK 下载完成后，出现如图 1.25 所示的界面，单击"Finish"按钮，完成 SDK 的下载和安装，此时弹出 Android Studio 主菜单界面，如图 1.26 所示。

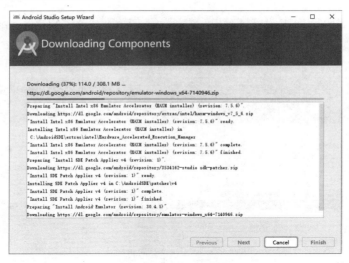

图 1.24　SDK 下载过程

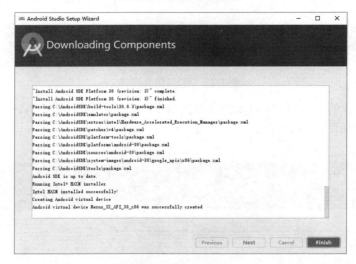

图 1.25　SDK 下载完成

图 1.26　Android Studio 主菜单

至此，Android Studio 及 SDK 已全部安装完成并启动。

本书安装 Android Studio 时使用的安装包是中文社区官网提供的 Android Studio 3.5.2 版本，安装成功并启动 Android Studio 后在线升级到 Android Studio 4.1.2 版本。当然也可以直接下载 Android Studio 4.1.2 版本进行安装。

3．启动 Android Studio 并创建 Android 工程

（1）在"开始"菜单中找到安装时创建的快捷方式，启动 Android Studio。第一次启动时默认创建工程，打开"Create New Project"对话框，如图 1.27 所示。

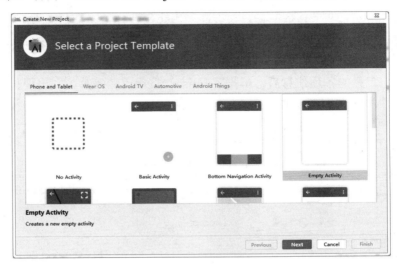

图 1.27　"Create New Project"对话框

（2）选择工程的类型，此处默认为"Empty Activity"类型，单击"Next"按钮进入"Configure Your Project"对话框，如图 1.28 所示。

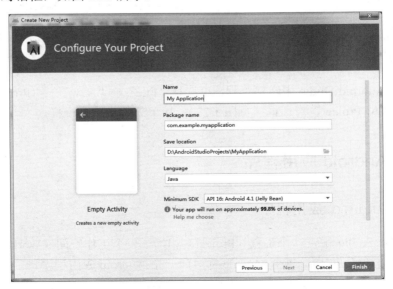

图 1.28　配置工程

（3）填写工程信息。

在"Name"编辑框中输入工程的名称，默认为"My Application"，可以修改。

在"Package name"编辑框中输入工程中 Java 程序所在的包名，默认为"com.example.myapplication"，可以修改。

在"Save location"编辑框中选择工程文件存放的路径，默认路径为"C:\Users\用户名\AndroidStudioProjects"（其中<用户名>是当前 Windows 系统的登录用户名）。此书中所有工程文件存放位置为"D:\AndroidStudioProjects"。

在"Language"下拉框中选择 Java。

在"Minimum SDK"下拉框中选择运行此应用程序目标设备的最低 API level，如果目标设备的 SDK 版本号低于这个版本，则无法进行安装。一般使用默认选项，因为高版本的 Android 会向下兼容，这样就能够兼容运行在所有手机上。

（4）填写完工程的各项信息后，单击"Finish"按钮，稍等片刻，出现如图 1.29 所示的 Android Studio 集成开发环境界面。

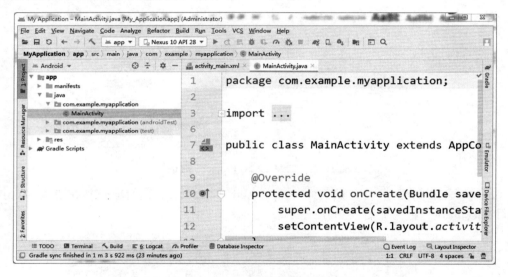

图 1.29　Android Studio 集成开发环境界面

如果要在 Android Studio 中新建工程，可选择"File"→"New"→"Project"菜单命令，打开"Create New Project"对话框（见图 1.27），接下来的步骤都一样。

1.3　开发 Android 应用程序

1.3.1　创建 Android 应用程序

在 Android Studio 中，一个工程（Project）相当于一个工作空间（Workspace），其中可以包括多个模块（Module），每个 Module 都对应一个 Android 应用程序。

在 Android Studio 的工程中新建一个 Module，其步骤如下。

(1)执行"File"→"New"→"Module"菜单命令,打开"Create New Module"对话框,如图 1.30 所示。

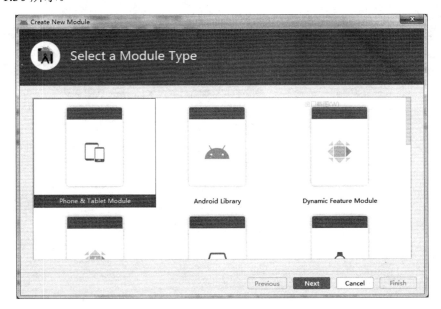

图 1.30　创建 Module

(2)选择 Module 的类型,默认为"Phone & Tablet Module",单击"Next"按钮,进入"Phone & Tablet Module"界面,如图 1.31 所示。

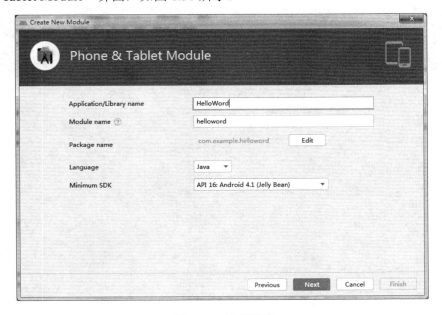

图 1.31　配置界面

(3)填写 Application/Library name,并设置 Language 和 Minimum SDK 选项,单击"Next"按钮,打开"Add an Activity to Mobile"界面,如图 1.32 所示。

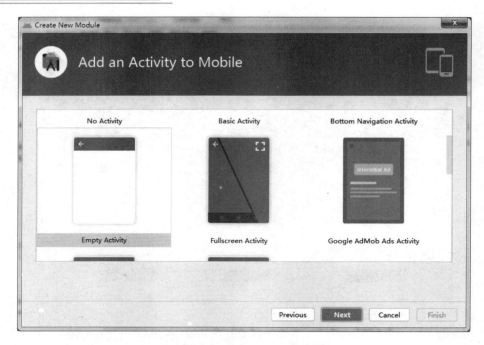

图 1.32　选择 Activity 类型

（4）选择"Empty Activity"选项，单击"Next"按钮，打开"Configure Activity"界面，如图 1.33 所示。

图 1.33　配置 Activity

（5）填写 Activity Name，默认勾选"Generate a Layout File"复选框，单击"Finish"按钮，完成 Module 的创建。

1.3.2 Android 应用程序目录结构

一个完整的 Android Studio 工程中包含很多的文件和资源，Android Studio 提供了 Project 工具窗口来辅助用户管理和查看工程的结构。Project 窗口位于开发界面的左上侧，有多种显示模式，如图 1.34 所示。

默认情况下，Android Studio 以 Android 模式展示工程中所有的包、目录和文件。例如，前面创建的"My Application"工程在 Project 工具窗口中的 Android 模式视图如图 1.35 所示。该工程中包含默认自动创建的 app 应用程序、helloworld 应用程序，以及工程架构信息 Gradle Scripts。

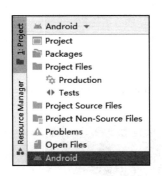

图 1.34 Project 工具窗口的显示模式

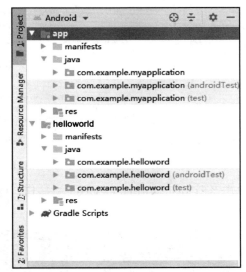

图 1.35 My Application 工程结构

一个典型的 Android Studio 应用程序的目录可分为三部分：manifests 目录、java 目录和 res 目录。

（1）manifests 目录：每一个应用程序都对应一个 manifests 类型的文件，其中 AndroidManifest.xml 存储应用程序的配置信息。

（2）java 目录：存放 Java 源代码的包。

（3）res 目录：存储项目的资源，包含以下子目录。

- drawable：存放应用程序所需的图片文件。
- layout：存放用户的布局（layout）文件。
- mipmap：存放应用程序的图片资源，可以按屏幕分辨率将图片分成 hdpi（高分辨率）、mdpi（低分辨率）等。
- values：存放用户界面使用的颜色、文字和样式等的定义文件。

1.3.3 创建 Android 模拟器

利用 Android Studio 自带仿真器（Android Virtual Device，AVD）来模拟智能手机，供开

发人员随时对智能手机进行运行和测试 App。在 Android Studio 中创建模拟器的步骤如下。

（1）执行"Tools"→"AVD Manager"菜单命令，或者单击工具栏中的 AVD Manager 工具，打开"Android Virtual Device Manager"窗口，如图 1.36 所示。

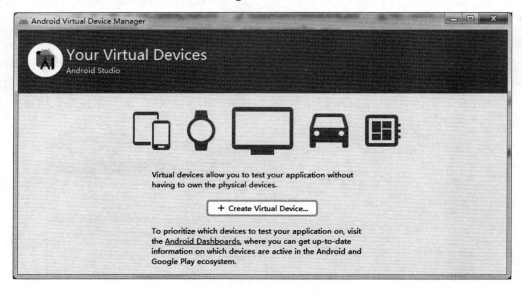

图 1.36　AVD 管理器

（2）单击"+ Create Virtual Device…"按钮，打开"Virtual Device Configuration"对话框，如图 1.37 所示。

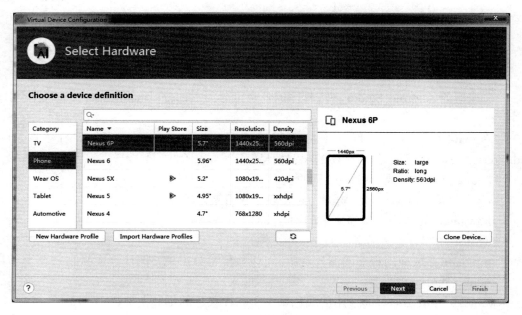

图 1.37　选择模拟设备

（3）选择模拟设备，单击"Next"按钮，打开如图 1.38 所示的系统镜像界面。

第 1 章 Android 入门

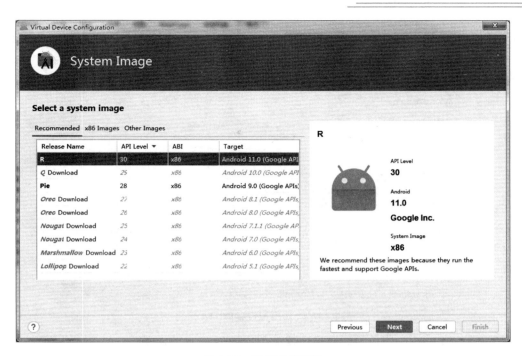

图 1.38 选择系统镜像

（4）选择模拟设备要安装的 Android 系统版本，如果没有下载相应的系统镜像，则选择"Download"选项，下载 SDK 文件并进行自动安装，如图 1.39 所示。

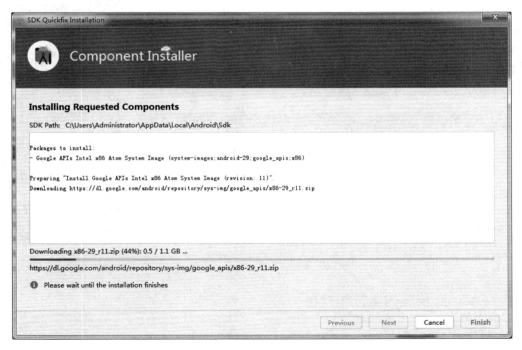

图 1.39 下载系统镜像

（5）安装完成后，单击"Finish"按钮，确认模拟设备的配置信息，如图 1.40 所示。

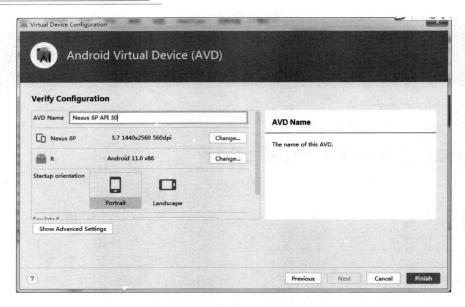

图 1.40 确认模拟设备的配置信息

（6）单击"Finish"按钮，完成模拟器的创建，如图 1.41 所示。

（7）选择模拟设备列表中的设备，单击"Actions"下方的启动按钮▶，启动模拟器，启动后的模拟器界面如图 1.42 所示。

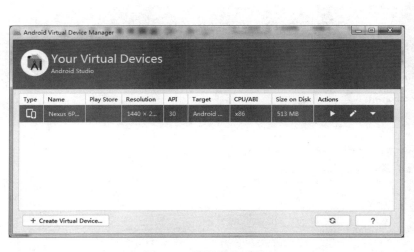

图 1.41 模拟设备列表

图 1.42 模拟器界面

1.3.4 Android 程序的运行和打包

1. 运行 Android 程序

模拟器创建并启动后，在工具栏中选择要运行的应用程序，单击运行按钮▶，如图 1.43 所示。稍等片刻就会出现一个手机屏幕界面，如图 1.44 所示。

第 1 章　Android 入门

图 1.43　运行程序　　　　　　图 1.44　手机屏幕界面

2．生成签名的 APK 文件

每个 Android 的 App 都需要一个证书对应用进行数字签名，否则无法安装。平时在手机上调试运行时，Android Studio 会自动用默认的密钥和证书来进行签名，但实际发布时，则需要进行手动签名。

生成签名的 APK 文件，其步骤如下。

（1）执行"Build"→"Generate Signed APK"菜单命令，打开"Generate Signed Bundle or APK"对话框，如图 1.45 所示。然后选中"APK"单选按钮，再单击"Next"按钮。

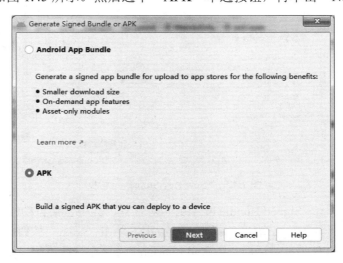

图 1.45　"Generate Signed Bundle or APK"对话框

（2）进入如图1.46所示对话框，单击"Create new…"按钮。

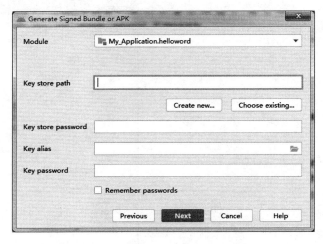

图1.46　单击"Create new…"按钮

（3）打开"New Key Store"对话框，设置签名文件命名、签名密码和证书内容，如图1.47所示。单击"OK"按钮，完成签名文件的创建。

图1.47　"New Key Store"对话框

（4）打开"Generate Signed Bundle or APK"对话框，其中包括自动输入的签名路径、签名路径密码、签名别名和签名密码，如图1.48所示，单击"Next"按钮。

（5）打开如图1.49所示对话框，在Build Variants选区中选择"release"选项，勾选"V2（Full APK Signature）Signature Help"复选框，单击"Finish"按钮生成签名的APK文件。

第 1 章　Android 入门

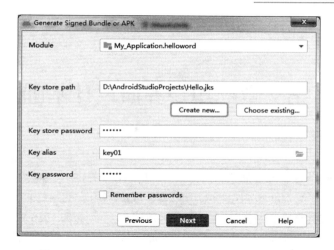

图 1.48　"Generate Signed Bundle or APK"对话框

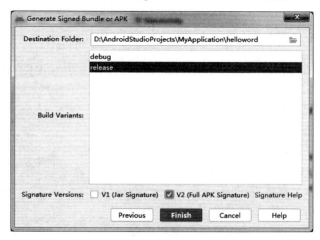

图 1.49　生成签名的 APK 文件

习题 1

1．Android 系统架构包括哪几个层次？各层有何特点？
2．简述 Android 的四大组件及其作用。
3．Android 开发环境的搭建。
（1）下载、安装和配置 JDK，创建 Android 模拟器，搭建 Android 的开发环境。
（2）编写一个 Android 程序，并在模拟器上运行，屏幕中央显示为"Hello Android!"。结合这个实例阐述 Android 应用程序的开发过程。

第 2 章　用户界面设计基础

用户使用应用程序时最先接触的就是应用程序的用户界面（User Interface，UI），用户界面设计是否美观，控件是否能够满足程序功能会直接影响用户的最终体验。Android SDK 为开发者提供了全方位的用户界面开发支持，包括完备的交互控件库、灵活的界面布局管理方案和灵活的事件监听机制。本章介绍 Android 用户界面设计、常用布局、常用基本控件等内容。

2.1　用户界面编写方式

在 Android 应用程序中用户使用控件和程序进行交互，所有控件在窗口内部有序排列，View 和 ViewGroup 是所有控件的直接或间接父类，其中，View 是绘制在屏幕上的（用户能与之交互）一个对象，ViewGroup 作为容器填充内部控件，并且应用程序界面的根元素必须是一个 ViewGroup 容器。ViewGroup 和 View 的关系如图 2.1 所示。

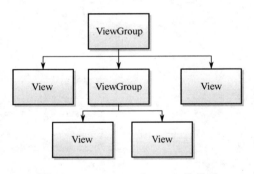

图 2.1　ViewGroup 和 View 的关系

Android SDK 为开发者提供了一个 View 和 ViewGroup 子类的集合，集合中提供了一些常用的输入控件和各种各样的布局模式，可以根据程序需要使用 Java 代码方式或在 XML 布局文件中编写布局。下面通过示例进行介绍。

【例 2.1】用户界面编写。

（1）使用 XML 布局文件创建 Android 界面。

首先在 Android Studio 集成开发环境中创建一个应用程序，并在 Android 目录结构中找到 res/layout 文件夹，打开 activity_xml.xml 布局文件，向其中添加一个 TextView 控件，布局文件的代码如下：

```
<?xml version="1.0" encoding="utf-8"?>
<RelativeLayout xmlns:android="http://schemas.android.com/apk/res/android"
    xmlns:app="http://schemas.android.com/apk/res-auto"
    xmlns:tools="http://schemas.android.com/tools"
    android:layout_width="match_parent"
    android:layout_height="match_parent"
    tools:context=".MainActivity">
    <TextView
        android:layout_width="wrap_content"
        android:layout_height="wrap_content"
```

```
            android:text="Hello World!"
            android:textColor="#ff0000"
            android:textSize="30dp"
            android:layout_centerInParent="true" />
</RelativeLayout>
```

该布局文件的根元素是一个 RelativeLayout 布局管理器，在其中定义了一个 TextView 节点，通过 textColor、textSize 和 text 属性设置文本框的相关显示属性，同时可以很方便地添加一些对齐属性。

编辑好布局文件后，打开应用程序的 MainActivity.java 文件，其中 onCreate()方法内的 setContentView(R.layout.activity_xml);代码负责加载布局文件并显示。使用 XML 布局文件创建界面如图 2.2 所示。

图 2.2　使用 XML 布局文件创建界面

（2）使用 Java 代码编写 Android 应用界面。

在 Android 中所有布局和控件的对象都可以通过 new 关键字创建出来，将创建的 View 控件添加到 ViewGroup 布局中，实现 View 控件在布局界面中的显示。

在上述应用程序的 MainActivity.java 文件的 onCreate()方法中编写以下代码：

```
public class MainActivity extends AppCompatActivity {
    @Override
    protected void onCreate(Bundle savedInstanceState) {
        super.onCreate(savedInstanceState);
        ConstraintLayout constraintLayout = new ConstraintLayout(this);
        ConstraintLayout.LayoutParams params = new ConstraintLayout.LayoutParams(
                ConstraintLayout.LayoutParams.WRAP_CONTENT,
                ConstraintLayout.LayoutParams.WRAP_CONTENT);
        TextView textView = new TextView(this);
        textView.setText("Hello Android!");
        textView.setTextColor(Color.BLACK);
        textView.setTextSize(30);
        constraintLayout.addView(textView, params);
        setContentView(constraintLayout);
    }
}
```

使用代码编写界面如图 2.3 所示。

在上述代码中，首先创建一个 ConstraintLayout 布局对象和 TextView 文本框对象，并通过 setText()、setTextColor()和 setTextSize()等方法设置文本框属性，然后使用 addView()方法将 TextView 对象添加到布局中，最后使用 setContentView()方法将 ConstraintLayout 布局显示在窗口中。

通过上述示例可以看出，开发 Android 界面可以使用 Java 代码和 XML 布局文件来达到开发目的。通常使用 XML 布局文件定义应用程序的界面框架，开发者根据程序逻辑需要在代码中动态改变控件的某些显示特征。总之，无论使用哪种方案编写界面其本质都是相同的，控制界面元素的 XML 属性也都有与其对应的方法。

图 2.3　使用代码编写界面

2.2　常用布局

针对不同的用户需求，Android SDK 提供了多种布局风格，分别是 LinearLayout、RelativeLayout、FrameLayout、TableLayout、GridLayout、ConstraintLayout 和 AbsoluteLayout 等，下面分别进行讲解。

2.2.1　布局通用属性

Android SDK 提供的布局类用于容纳用户控件，通过属性设置控件的摆放方式、相对位置和大小，所有的基本布局类都可直接或间接继承于 ViewGroup 类，布局通用属性如表 2.1 所示。

表 2.1　布局通用属性

属　　性	描　　述
xmlns:android xmlns:app xmlns:tools	命名空间
android:id	通过"@+id/..."设置布局 id
android:layout_width	设置布局宽度，取值为 wrap_content 或 match_parent，分别代表由填充内容决定或与父控件相同
android:layout_height	设置布局高度，取值为 wrap_content 或 match_parent，分别代表由填充内容决定或与父控件相同
android:padding	内边距（内部空间和布局边缘的间距）

2.2.2　LinearLayout

LinearLayout（线性布局）指在界面内按水平或垂直方向摆放控件，在 XML 布局文件中的声明方式为：

```
<LinearLayout xmlns:android="http://schemas.android.com/apk/res/android"
    xmlns:app="http://schemas.android.com/apk/res-auto"
    xmlns:tools="http://schemas.android.com/tools">
```

```
        ...
    </LinearLayout>
```

LinearLayout 通过 android:orientation 属性设置其内部控件的摆放方式，值为"horizontal"时表示水平摆放，值为"vertical"时表示垂直摆放。

【例 2.2】LinearLayout 示例。

在 Android Studio 集成开发环境中创建一个应用程序，打开布局文件 activity_layout.xml，向其中添加三个 TextView 控件。

activity_layout.xml 布局文件代码如下：

```xml
<?xml version="1.0" encoding="utf-8"?>
<LinearLayout
    xmlns:android="http://schemas.android.com/apk/res/android"
    xmlns:app="http://schemas.android.com/apk/res-auto"
    xmlns:tools="http://schemas.android.com/tools"
    android:id="@+id/l1"
    android:layout_width="match_parent"
    android:layout_height="match_parent"
    android:orientation="vertical "
    tools:context=". MainActivity ">
    <TextView
        android:id="@+id/textView2"
        android:layout_width="wrap_content"
        android:layout_height="wrap_content"
        android:text="TextView1"
        android:textColor="@color/black"
        android:textSize="24sp" />
    <TextView
        android:id="@+id/textView3"
        android:layout_width="wrap_content"
        android:layout_height="wrap_content"
        android:text="TextView2"
        android:textColor="@color/purple_200"
        android:textSize="24sp" />
    <TextView
        android:id="@+id/textView4"
        android:layout_width="wrap_content"
        android:layout_height="wrap_content"
        android:text="TextView3"
        android:textColor="@color/teal_200"
        android:textSize="24sp" />
</LinearLayout>
```

垂直线性布局如图 2.4 所示，将"android:orientation"属性设置为"horizontal"后，水平线性布局如图 2.5 所示。

在图 2.5 中三个水平摆放的 TextView 控件首尾相连很不美观，这时可以通过"android:layout_weight"属性设置每个 TextView 控件在线性布局水平方向上所占的比例。

使用 weight 属性需要先将每个 TextView 控件的"android:layout_width"属性设置为"0dp"，同时添加代码：android:layout_weight="1"，weight 属性显示结果如图 2.6 所示。

上述代码中，当三个 TextView 控件的属性"android:layout_weight"都是"1"时，表示这三个控件的宽度比为"1：1：1"，所以程序的运行结果是在同一行内三个控件为等宽度显示,要实现其他的显示效果时可灵活调整权值。

为了使界面更加美观,可以在控件之间添加分割线。在布局文件 activity_layout.xml 中的两个 TextView 控件之间,添加如下代码即可实现:

```xml
<View
    android:layout_width="match_parent"
    android:layout_height="1dp"
    android:background="#000000" />
```

同时将"android:orientation"属性修改为"vertical",显示结果如图 2.7 所示。通过在两个 TextView 控件之间添加一个 View 控件,设置 width 属性和 height 属性使其显示为一条水平分割线。

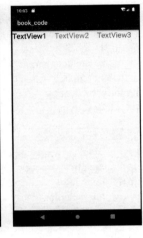

图 2.4　垂直线性布局　　图 2.5　水平线性布局　　图 2.6　weight 属性　　图 2.7　分割线

2.2.3　RelativeLayout

RelativeLayout(相对布局)是指通过相对定位的方式指定容器内部控件的位置,在 XML 布局文件中的声明方式为:

```xml
<RelativeLayout xmlns:android="http://schemas.android.com/apk/res/android"
    xmlns:app="http://schemas.android.com/apk/res-auto"
    xmlns:tools="http://schemas.android.com/tools">
    ...
</RelativeLayout>
```

RelativeLayout 可以使用相关属性通过父容器或通过兄弟组件进行定位。在相对布局容器中,当 A 组件的位置由 B 组件决定时,Android 要求先定义 B 组件,再定义 A 组件。

1. 根据父容器定位

相对容器内部控件可以设定相对父容器边界进行位置定位,如图 2.8 所示。

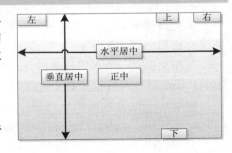

图 2.8　相对父容器定位

RelativeLayout 提供一个内部类——RelativeLayout.LayoutParams 所提供的大量 XML 属性来控制子控件的位置摆放，通过设置属性值为"true"或"false"来确定其和父容器之间的位置关系，父容器定位属性如表 2.2 所示。

表 2.2 父容器定位属性

属　　性	描　　述	属　　性	描　　述
Android:layout_centerHorizontal	水平居中	android:layout_alignParentLeft	贴紧父控件的左边缘
android:layout_centerVertical	垂直居中	android:layout_alignParentRight	贴紧父控件的右边缘
android:layout_centerInparent	相对于父控件完全居中	android:layout_alignParentTop	贴紧父控件的上边缘
android:ayout_alignParentBottom	贴紧父控件的下边缘		

2．根据兄弟控件定位

兄弟控件指处于同一层次容器内的控件。RelativeLayout 提供了一组根据与兄弟控件的位置关系来确定自身位置的属性，这些属性值为其兄弟控件的 id，其属性如表 2.3 所示。

表 2.3 兄弟控件定位属性

属　　性	描　　述
Android:layout_below	在某控件下方
android:layout_above	在某控件上方
android:layout_toLeftOf	在某控件的左边
android:layout_toRightOf	在某控件的右边
android:layout_alignTop	本控件的上边缘和某控件的上边缘对齐
android:layout_alignLeft	本控件的左边缘和某控件的左边缘对齐
android:layout_alignBottom	本控件的下边缘和某控件的下边缘对齐
android:layout_alignRight	本控件的右边缘和某控件的右边缘对齐

【例 2.3】RelativeLayout 示例。

在 Android Studio 中创建应用程序，编写布局文件 activity_layout.xml，代码如下：

```xml
<?xml version="1.0" encoding="utf-8"?>
<RelativeLayout
    xmlns:android="http://schemas.android.com/apk/res/android"
    xmlns:app="http://schemas.android.com/apk/res-auto"
    xmlns:tools="http://schemas.android.com/tools"
    android:layout_width="match_parent"
    android:layout_height="match_parent"
    tools:context=".MainActivity">
    <TextView
        android:id="@+id/c"
        android:layout_width="wrap_content"
        android:layout_height="wrap_content"
        android:layout_centerInParent="true"
        android:text="中"
        android:textSize="60sp"
        android:textStyle="bold" />
    <TextView
```

```
            android:id="@+id/n"
            android:layout_width="wrap_content"
            android:layout_height="wrap_content"
            android:layout_above="@id/c"
            android:layout_centerInParent="true"
            android:text="北"
            android:textSize="60sp"
            android:textStyle="bold" />
    <TextView
            android:id="@+id/s"
            android:layout_width="wrap_content"
            android:layout_height="wrap_content"
            android:layout_below="@id/c"
            android:layout_centerInParent="true"
            android:text="南"
            android:textSize="60sp"
            android:textStyle="bold" />
    <TextView
            android:id="@+id/w"
            android:layout_width="wrap_content"
            android:layout_height="wrap_content"
            android:layout_centerInParent="true"
            android:layout_toLeftOf="@id/c"
            android:text="西"
            android:textSize="60sp"
android:textStyle="bold" />
    <TextView
            android:id="@+id/e"
            android:layout_width="wrap_content"
            android:layout_height="wrap_content"
            android:layout_centerInParent="true"
            android:layout_toRightOf="@id/c"
            android:text="东"
            android:textSize="60sp"
            android:textStyle="bold" />
</RelativeLayout>
```

RelativeLayout 示例运行结果如图 2.9 所示。

在上述代码中，XML 布局文件使用 RelativeLayout 作为根元素，在其中添加五个 TextView 子控件，首先相对于父容器居中摆放一个控件，其他分别作为兄弟控件相对其位置上、下、左、右摆放以达成最终的显示效果。

2.2.4 FrameLayout

FrameLayout（帧布局）指在屏幕上开辟出一块空白的区域，当往里面添加控件时，会默认把它们放到这块区域的左上角，并且每个子控件独占一帧。帧布局的大小由控件中最大的子控件决定，如果控件的大小一样的话，那么同一时刻就只能看到最上面的那个控件，后续添加的控件

图 2.9　RelativeLayout 示例运行结果

会覆盖前一个。在 XML 布局文件中 FrameLayout 的声明方式为：
```
<FrameLayout xmlns:android="http://schemas.android.com/apk/res/android"
    xmlns:app="http://schemas.android.com/apk/res-auto"
    xmlns:tools="http://schemas.android.com/tools">
    ...
</FrameLayout>
```

FrameLayout 可以通过元素属性设置前景图像，前景图像永远处于帧布局最上面而不会被其他控件覆盖。其相关属性为：

android:foreground：设置帧布局容器的前景图像。

android:foregroundGravity：设置前景图像显示的位置。

【例 2.4】FrameLayout 示例。

创建应用程序，编写布局文件 activity_layout.xml，代码如下：

```xml
<?xml version="1.0" encoding="utf-8"?>
<FrameLayout
    xmlns:android="http://schemas.android.com/apk/res/android"
    xmlns:app="http://schemas.android.com/apk/res-auto"
    xmlns:tools="http://schemas.android.com/tools"
    android:layout_width="match_parent"
    android:layout_height="match_parent"
    android:foreground="@mipmap/ic_launcher"
    android:foregroundGravity="right|bottom"
    tools:context=".MainActivity">
    <TextView
        android:layout_width="200dp"
        android:layout_height="200dp"
        android:background="#FF6143" />
    <TextView
        android:layout_width="150dp"
        android:layout_height="150dp"
        android:background="#7BFE00" />
    <TextView
        android:layout_width="100dp"
        android:layout_height="100dp"
        android:background="#FFFF00" />
</FrameLayout>
```

FrameLayout 示例运行结果如图 2.10 所示。

在上述代码中，XML 布局文件以 FrameLayout 为根元素，在其中设置三个大小和颜色不同的 TextView 控件，可以发现每个控件都占用一帧，后定义的控件放置在前端。图像作为前景时一定要放置在所有控件之前。

2.2.5 TableLayout

TableLayout（表格布局）指通过在其中添加 TableRow 来控制布局的行数，并以添加 TabeRow 中控件的个数来控制表格的列数。

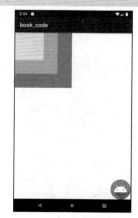

图 2.10　FrameLayout 示例的运行结果

在 XML 布局文件中声明 TableLayout 的基本代码为：
```xml
<TableLayout xmlns:android="http://schemas.android.com/apk/res/android"
    xmlns:app="http://schemas.android.com/apk/res-auto"
    xmlns:tools="http://schemas.android.com/tools">
    ...
</TableLayout>
```

在 TableLayout 中，一个 TableRow 代表一行，表格布局中有多少个 TableRow 就有多少行。TableRow 中控件的个数决定了表格布局的列数，也就是表格布局的列数由包含子控件最多的那一行决定。整个表格布局的宽度取决于父容器的宽度，列的宽度由本列中最宽的控件决定。

TableLayout 可以控制子控件摆放的列是否跨列，以及列的一些特殊显示效果，如拉伸、收缩和隐藏等，其常用属性如表 2.4 所示。

表 2.4 TableLayout 的常用属性

属 性	描 述	属 性	描 述
android:layout_column	控件显示的位置	android:shrinkColumns	设置允许被收缩列的列序号
android:layout_span	控件占用的列数	android:stretchColumns	设置运行被拉伸列的列序号
android:collapseColumns	设置需要被隐藏列的序号		

需要注意的是，在 TableLayout 中列号从"0"开始计算。

【例 2.5】TableLayout 示例。

创建应用程序，编写布局文件 activity_layout.xml，代码如下：

```xml
<?xml version="1.0" encoding="utf-8"?>
<TableLayout xmlns:android="http://schemas.android.com/apk/res/android"
    xmlns:app="http://schemas.android.com/apk/res-auto"
    xmlns:tools="http://schemas.android.com/tools"
    android:layout_width="match_parent"
    android:layout_height="match_parent"
    android:collapseColumns="0,2"
    android:shrinkColumns="3"
    android:stretchColumns="1"
    tools:context=".MainActivity">
    <TableRow>
        <Button
            android:layout_width="wrap_content"
            android:layout_height="wrap_content"
            android:text="one" />
        <Button
            android:layout_width="wrap_content"
            android:layout_height="wrap_content"
            android:text="two" />
        <Button
            android:layout_width="wrap_content"
            android:layout_height="wrap_content"
            android:text="three" />
        <Button
            android:layout_width="wrap_content"
            android:layout_height="wrap_content"
            android:text="four" />
        <Button
```

```
                    android:layout_width="wrap_content"
                    android:layout_height="wrap_content"
                    android:text="five" />
            </TableRow>
            <TableRow>
                <Button
                    android:layout_column="3"
                    android:layout_width="wrap_content"
                    android:layout_height="wrap_content"
                    android:text="one" />
            </TableRow>
</TableLayout>
```

在上述代码中，XML 布局文件采用 TableLayout 为根元素，同时设置第 0 列和第 2 列为隐藏类，第 1 列为扩张列，在表格布局中添加两行，每行都使用按钮作为子控件，可以发现按钮的摆放和大小都发生了改变。

TableLayout 示例运行结果如图 2.11 所示。

2.2.6 GridLayout

GridLayout（网格布局）是 Android 4.0 以后引入的一个新布局，初学者往往分不清它和 TableLayout 的区别，使用 GridLayout 可以减少布局嵌套，同时使用新增的布局属性可以细粒度地控制子控件位置，GridLayout 的常用属性如表 2.5 所示。

图 2.11　TableLayout 示例的运行结果

表 2.5　GridLayout 的常用属性

属　　性	描　　述
android:columnCount	最大列数
android:rowCount	最大行数
android:orientation	GridLayout 中子元素的布局方向
android:alignmentMode	alignBounds 指对齐子视图边界；alignMargins 指对齐子视图内容，默认值
android:columnOrderPreserved	使列边界显示的顺序和列索引的顺序相同，默认值是 true
android:rowOrderPreserved	使行边界显示的顺序和行索引的顺序相同，默认值是 true
android:useDefaultMargins	没有指定视图的布局参数时，使用默认的边距，默认值是 false

在 XML 布局文件中声明 GridLayout 的基本代码为：

```
<GridLayout xmlns:android="http://schemas.android.com/apk/res/android"
    xmlns:app="http://schemas.android.com/apk/res-auto"
    xmlns:tools="http://schemas.android.com/tools">
    ...
</GridLayout>
```

【例 2.6】GridLayout 示例。

创建应用程序，编写布局文件 activity_layout.xml，代码如下：

```xml
<?xml version="1.0" encoding="utf-8"?>
<GridLayout xmlns:android="http://schemas.android.com/apk/res/android"
    xmlns:app="http://schemas.android.com/apk/res-auto"
    xmlns:tools="http://schemas.android.com/tools"
    android:layout_width="match_parent"
    android:layout_height="match_parent"
    android:columnCount="2"
    android:orientation="horizontal"
    android:rowCount="2"
    tools:context=".MainActivity">
    <Button
        android:id="@+id/one"
        android:layout_column="1"
        android:text="1" />
    <Button
        android:id="@+id/two"
        android:layout_column="0"
        android:text="2" />
    <Button
        android:id="@+id/three"
        android:text="3" />
    <Button
        android:id="@+id/devide"
        android:text="/" />
</GridLayout>
```

在上述代码中，XML 布局文件声明根元素为 GridLayout，使用 android:rowCount 和 android:columnCount 的属性设置网格行数和列数，然后在其中添加一些子控件，GridLayout 示例运行结果如图 2.12 所示。

2.2.7　ConstraintLayout

ConstraintLayout（约束布局）是 Android Studio 2.2 中主要的新增功能之一，也是谷歌公司在 2016 年的 I/O 大会上重点宣传的一个功能。它的出现解决了布局嵌套过多的问题，可采用灵活的方式定位和调整小部件。

在传统的 Android 开发中，界面都是靠编写 XML 代码完成的，ConstraintLayout 非常适合使用可视化的方式来编写界面，当然，可视化操作的背后仍然是使用 XML 代码来实现的，只不过这些代码是由 Android Studio 根据用户操作自动生成的，极大地提升了 Android 界面的开发效率。

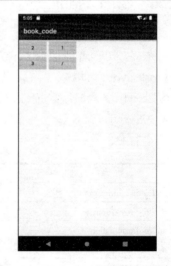

图 2.12　GridLayout 示例的运行结果

在 Android Studio 中创建一个应用程序，打开 activity_layout.xml 文件的"design"模式，如图 2.13 所示。

可以发现，Android Studio 默认的布局方式就是 ConstraintLayout，此时可以通过拖曳控件、添加约束等可视化方式进行界面设计。

1．基本操作

从左侧 Palette 面板向设计窗口内拖入一个按钮，此时按钮上没有施加任何约束，所以在预览窗口内按钮的位置并不是程序运行时的位置，按钮边框上的小圆圈就是添加约束的位置，如图 2.14 所示。

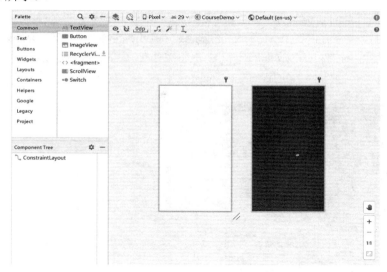

图 2.13　ConstraintLayout 编辑面板

控件约束分为垂直和水平两类，可以在四个方向上给控件添加约束，通过鼠标拖曳操作向按钮四周添加四个约束，如图 2.15 所示，按钮显示位置固定在容器正中。

约束不仅可以在控件与容器之间添加，还可以使用约束让一个控件相对于另一个控件进行定位。例如，在设计窗口中添加一个 Button 控件，让其位于第一个 Button 控件的正下方，间距为 50dp，要达到此目的就需要在第二个按钮中添加相对于第一个按钮的约束，如图 2.16 所示。

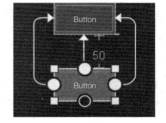

图 2.14　约束点　　　　　图 2.15　简单约束　　　　　图 2.16　控件间相对约束

在进行复杂界面设计时，简化不同控件之间的对齐操作可以借助于 GudieLine 工具。参考线 GuideLine 可用于 Views 的对齐参照，而且不会在运行时显示。

当前的参考线 GuideLine 有三种类型，默认的第一种参考线会有一个固定的偏移向父组件的 start 边缘（偏移量的单位是 dp）。第二种参考线有一个固定的偏移向父组件的 end 边缘。第三种参考线是根据父组件 ConstraintLayout 的宽度百分比来放置的。通过单击设计视图

GuideLine 边缘的标识按钮来切换参考线的类型。图 2.17 演示了在设计窗口中添加一条水平 GuideLine，切换类型为百分比型，然后拖曳到窗口中央（50%），即分别在上半部分和下半部分中添加按钮控件，并添加相对容器和相对 GuideLine 的约束，这样就对齐子控件了。

通过选择约束子控件的摆放来确定位置，还需要使用 Inspector 设计控件的其他显示特性。当选中任意一个控件时，在右侧的 Properties 区域会出现很多属性选项，如图 2.18 所示。

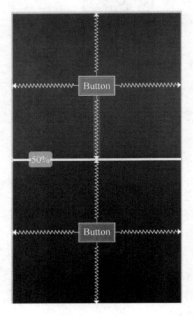 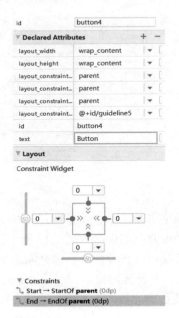

图 2.17 GuideLine   图 2.18 Properties 区域

在 Properties 区域可以设置当前控件的所有属性，如文本内容、颜色、点击事件等。Properties 区域的下半部分被称为 Inspector。在 Inspector 中有一个纵向的轴和一个横向的轴，这两个轴也是用于确定控件位置的，如果调整这两个轴的比例，那么 Button 的位置也会随之改变。位于 Inspector 中间的那个正方形区域，是用来控制控件大小的。它有 3 种模式可选，每种模式都使用一种不同的符号表示，单击符号即可进行切换。其中，<<表示 wrap content；⊢⊣表示固定值；↦↤表示 any size。any size 同 match parent 的区别是，match parent 用于填充当前控件的父布局，而 any size 用于填充当前控件的约束规则。

2．自动添加约束

在复杂界面设计中，为每个控件逐个添加约束非常烦琐。为此，ConstraintLayout 支持自动添加约束的功能，可以最大限度地简化操作。

自动添加约束的方式主要有两种，即 Autoconnect 和 Inference。使用 Autoconnect 就需要先在工具栏中单击 图标，将这个功能启用，默认情况下 Autoconnect 是不启用的，此时 Autoconnect 可以根据拖曳控件的状态自动判断该如何添加约束。不过 Autoconnect 是无法保证能百分百做出准确判断的。如果自动添加的约束并不是想要的，开发人员还可以随时进行手动修改。

Inference 也是用于自动添加约束的，但它比 Autoconnect 的功能更为强大。因为 Autoconnect 只能给当前操作的控件自动添加约束，而 Inference 可以给当前界面中的所有元素自动添加约束。

因而 Inference 适合实现复杂度比较高的界面，它可以一键自动生成所有的约束。例如，需要设计一个登录界面，首先将需要的控件拖入设计窗口中，如图 2.19 所示。然后，将各个控件按照界面设计的位置进行摆放，最后单击工具栏上的 Infer Constraints 按钮 ，就能为所有控件自动添加约束了，如图 2.20 所示。

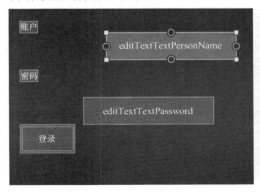

图 2.19　添加约束前的控件　　　　　图 2.20　使用 Inference 添加约束

3．链

如果两个或以上控件通过图 2.21 所示的方式约束在一起，就可以认为是一条链，图中为横向的链，纵向的链同理。

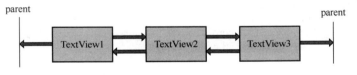

图 2.21　链

在一条链中，第一个控件是链头，可以设置 layout_constraintHorizontal_ chainStyle 来改变整条链的样式。chains 提供如下三种样式。

CHAIN_SPREAD：展开元素（默认）。

CHAIN_SPREAD_INSIDE：展开元素，但链的两端贴近 parent。

CHAIN_PACKED：链的元素将被打包在一起。

2.2.8　AbsoluteLayout

AbsoluteLayout（绝对布局）使用 layout_x 和 layout_y 属性直接定义控件的位置、layout_width 和 layout_height 属性决定控件的大小，这种布局方式看似简单，但是在不同尺寸和分辨率的设备上显示效果不同，其适配性差，所以在实际开发中几乎不会使用，在这里也就不展开讲解了。

2.3　常用控件

Android SDK 提供了非常丰富的控件，开发人员使用这些控件可以很方便地进行用户界面的开发。

2.3.1 TextView 控件

TextView（文本视图）控件用于显示文本，其常用属性如表 2.6 所示。

表 2.6 TextView 控件的常用属性

属 性	描 述
id	TextView 控件设置一个组件 id
layout_width	组件的宽度（wrap_content/match_parent）
layout_height	组件的高度
gravity	设置控件中内容的对齐方向
text	设置显示的文本内容，一般把字符串输入 string.xml 文件中，并通过 "@String/***" 取得对应的字符串内容
textColor	设置字体颜色，可通过 colors.xml 资源来引用
textStyle	设置字体风格（normal/bold/talic）
textSize	字体大小，单位一般为 sp
background	控件的背景颜色或图片

【例 2.7】TextView 示例。

创建一个应用程序，编辑 XML 布局文件，使用 LinearLayout，向其中添加一个 TextView 控件，代码如下：

```xml
<?xml version="1.0" encoding="utf-8"?>
<LinearLayout
xmlns:android="http://schemas.android.com/apk/res/android"
    xmlns:app="http://schemas.android.com/apk/res-auto"
    xmlns:tools="http://schemas.android.com/tools"
    android:layout_width="match_parent"
    android:layout_height="match_parent"
    android:background="#8fffad"
    android:gravity="center"
    tools:context=".MaiActivity">
    <TextView
        android:id="@+id/txtOne"
        android:layout_width="200dp"
        android:layout_height="200dp"
        android:background="#000000"
        android:gravity="center"
        android:text="TextView 示例(显示框)"
        android:textColor="#EA5246"
        android:textSize="18sp"
        android:textStyle="bold|italic" />
</LinearLayout>
```

TextView 示例运行结果如图 2.22 所示。

Android 为 TextView 控件提供了设计文字阴影的属性，具体内容如下。

android:shadowColor：设置阴影的颜色。

android:shadowRadius：设置阴影的模糊程度。

android:shadowDx：设置阴影在水平方向的偏移，即水平方向阴影开始的横坐标位置。

android:shadowDy：设置阴影在垂直方向的偏移，即垂直方向阴影开始的纵坐标位置。

在上例 TextView 控件中添加如下属性。

```
android:shadowColor="#F9F900"
android:shadowDx="10.0"
android:shadowDy="10.0"
android:shadowRadius="3.0"
```

带阴影的 TextView 如图 2.23 所示。

图 2.22　TextView 示例运行结果　　　　图 2.23　带阴影的 TextView

2.3.2　EditText 控件

EditText（文本编辑框）控件可以接收用户输入，默认多行显示，并且能够自动换行，即当一行显示不完全时，会自动换到第二行。EditText 控件是 TextView 的子类，其常用属性如表 2.7 所示。

表 2.7　EditText 控件的常用属性

属　　性	描　　述
android:hint	提示文本
android:textColorHint	设置提示信息的颜色
android:cursorVisible	设定光标为显示或隐藏，默认显示
android:drawableLeft	在 Text 的左边输出一个 drawable
android:editable	设置是否可编辑
android:ems	设置宽度为 N 个字符的宽度
android:maxLength	限制显示的文本长度，超出部分不显示
android:lines	设置文本的行数
android:scrollHorizontally	当设置文本超出 TextView 的宽度时，是否出现滚动条

EditText 控件使用 android:inputType 属性设置输入类型，其内容如下。

Number：只能输入数字。

numberDecimal：只能输入浮点数（小数）整数。

Password：将输入的文字显示为"***"，用于用户输入密码。

Phone：拨号键盘。

Text：文本格式。

EditText 控件通过 addTextChangedListener()方法添加文本改变监听器。监听器需要实现的方法如下。

beforeTextChanged()：文本改变前调用。

onTextChanged()：文本改变中调用。

afterTextChanged()：文本改变后调用。

【例 2.8】EditText 示例。

创建应用程序，编辑 XML 布局文件，使用 LinearLayout，向其中添加一个 EditText 控件，代码如下：

```xml
<?xml version="1.0" encoding="utf-8"?>
<LinearLayout
    xmlns:android="http://schemas.android.com/apk/res/android"
    xmlns:app="http://schemas.android.com/apk/res-auto"
    xmlns:tools="http://schemas.android.com/tools"
    android:layout_width="match_parent"
    android:layout_height="match_parent"
    android:orientation="vertical"
    tools:context=".MainActivity">
    <EditText
        android:id="@+id/pwdText"
        android:layout_width="match_parent"
        android:layout_height="0dp"
        android:layout_weight="1"
        android:ems="10"
        android:gravity="center_vertical"
        android:hint="请输入登录密码"
        android:inputType="textPassword"
        android:selectAllOnFocus="true"
        android:textColorHint="#FF999999"
        android:textSize="27sp" />
</LinearLayout>
```

EditText 示例运行结果如图 2.24 所示。

图 2.24　EditText 示例运行结果

2.3.3 Button 控件

Button（按钮）控件继承于 TextView 控件，既可以显示文本，也可以通过设置背景属性显示图像。Button 控件允许用户通过单击来执行操作，在发生单击动作时，Button 控件的背景会同时发生切换操作，如同按钮被按下一样。

【例 2.9】Button 示例。

（1）界面设计。

创建应用程序，编辑 XML 布局文件，使用 LinearLayout 线性水平布局，向其中添加一个 TextView 控件和两个 Button 控件。布局文件代码如下：

```xml
<?xml version="1.0" encoding="utf-8"?>
<LinearLayout
    xmlns:android="http://schemas.android.com/apk/res/android"
    xmlns:app="http://schemas.android.com/apk/res-auto"
    xmlns:tools="http://schemas.android.com/tools"
    android:layout_width="match_parent"
    android:layout_height="match_parent"
    tools:context=".MainActivity">
    <Button
        android:id="@+id/add"
        android:layout_width="wrap_content"
        android:layout_height="wrap_content"
        android:layout_marginTop="20dp"
        android:layout_weight="1"
        android:text="Add"
        android:textSize="24sp" />
    <TextView
        android:id="@+id/textViewNum"
        android:layout_width="wrap_content"
        android:layout_height="wrap_content"
        android:layout_marginTop="20dp"
        android:layout_weight="1"
        android:text="0"
        android:textAlignment="center"
        android:textSize="24sp" />
    <Button
        android:id="@+id/sub"
        android:layout_width="wrap_content"
        android:layout_height="wrap_content"
        android:layout_marginTop="20dp"
        android:layout_weight="1"
        android:text="Sub"
        android:textSize="24sp" />
</LinearLayout>
```

（2）编写按钮的单击事件处理程序。

单击"Add"按钮时 TextView 控件中的数字会加 1，单击"Sub"按钮时 TextView 控件中的数字会减 1，此时需要使用 Button 控件类的 setOnClickListener()方法为按钮添加单击事件监听器，该监听器实现了 View.OnClickListener 接口，需要重写 onClick()方法。此方法在按钮控件被单击时调用。

修改 MainActivity 类代码，先通过添加私有属性"num"来记录 TextView 控件的显示内容，代码如下：

```java
private int num = 0;
```

修改 onCreate()方法代码如下：

```java
@Override
protected void onCreate(Bundle savedInstanceState) {
    super.onCreate(savedInstanceState);
    setContentView(R.layout.activity_button);
    TextView numTextView = (TextView) findViewById(R.id.textViewNum);
    Button addBtn = (Button) findViewById(R.id.add);
    Button subBtn = (Button) findViewById(R.id.sub);
    addBtn.setOnClickListener(new View.OnClickListener() {
        @Override
        public void onClick(View view) {
            num++;
            numTextView.setText(num);
        }
    });
    subBtn.setOnClickListener(new View.OnClickListener() {
        @Override
        public void onClick(View view) {
            num--;
            numTextView.setText(num);
        }
    });
}
```

Button 示例运行结果如图 2.25 所示。

上述代码中，先通过 findViewById()方法根据控件 id 初始化控件对象，然后为按钮控件注册事件监听器，并重写事件处理方法。

此时，运行程序就会发现单击按钮后，TextView 控件中的显示内容发生了变化，这说明按钮控件上的事件监听器起作用了。本示例里面的事件监听器使用了匿名内部类，除此之外，要为按钮设置单击事件处理还可以采用以下两种方法。

① 使用 onClick 属性。

在 XML 布局文件中指定 onClick 属性，其属性值为相应 Activity 类文件中定义的事件处理方法名，示例代码如下：

```xml
<Button
    ...
    Android:onClick="click"/>
```

click 方法的声明为：

```java
public void click(View view){
    ...
}
```

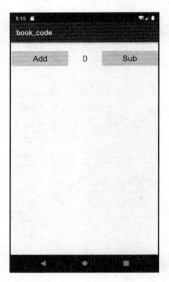

图 2.25 Button 示例运行结果

② Activity 类实现 View.OnClickListener 接口。先实现 onClick()方法，然后调用 Button 控件对象的 setOnClickListener()方法注册事件监听器，该方法的参数可以直接为 this。

综上所述，Android 处理 Button 控件相应的方法有，使用匿名内部类、onClick 属性定义和声明实现 View.OnClickListener 接口的类，这几种方法效果完全相同，前两种方法适用于控件较少的情况，当控件数目较多时推荐使用最后一种方法。同时，匿名内部类事件监听器适用于单击事件逻辑比较简单的情况。

Android SDK 还提供了 ImageButton 控件。它是一个图像背景的按钮，在使用上和 Button 控件类似，在这里就不再展开讲解了。

2.3.4　ImageView 控件

ImageView（图片视图）控件继承于 View 组件，是用来显示图片的一个控件。实际上它不仅可以用来显示图片，而且任何 Drawable 对象都可以使用 ImageView 来显示，并且 Android 为其提供了缩放和着色操作。ImageView 控件的常用属性如表 2.8 所示。

表 2.8　ImageView 控件的常用属性

属　性	描　述	属　性	描　述
android:maxHeight	设置 ImageView 的最大高度	android:background	背景
android:maxWidth	设置 ImageView 的最大宽度	android:src	图片内容
android:scaleType	设置所显示的图片如何缩放或移动以适应 ImageView 的大小	android:foreground	前景

ImageView 控件可以设置图片的属性，分别是 src、background 和 foreground，其中 background 表示背景，foreground 表示前景，src 表示内容，三者可以同时使用。src 填入图片时只能按照图片大小直接填充，并不会进行拉伸，而 background 和 foreground 填入图片时则会根据 ImageView 给定的宽度进行拉伸。background 和 foreground 是所有 View 控件都有的属性，能缩放 View 的大小，不受 scaleType 影响。src 是 ImageView 特有的属性，它会受到 scaleType 的影响。同时，虽然 foreground 和 src 都是前景，但是 foreground 在 src 之上，从层级上比较为 foreground>src>background。

当图片和 ImageView 控件的大小不同时，使用 scaleType 会直接影响图片的显示效果，scaleType 属性的取值如下。

matrix：使用 matrix 方式进行缩放。

fitXY：横向、纵向独立缩放，以适应该 ImageView。

fitStart：保持纵横比缩放图片，并且将图片放在 ImageView 的左上角。

fitCenter：保持纵横比缩放图片，缩放完成后将图片放在 ImageView 的中央。

fitEnd：保持纵横比缩放图片，缩放完成后将图片放在 ImageView 的右下角。

center：把图片放在 ImageView 的中央，但不进行任何缩放。

centerCrop：保持纵横比缩放图片，以使图片能完全覆盖 ImageView。

centerInside：保持纵横比缩放图片，以使 ImageView 能完全显示该图片。

【例 2.10】ImageView 示例。

（1）创建应用程序，复制图片 a.jpeg 到 res/drawable 目录中。

（2）编辑 XML 布局文件，使用 LinearLayout 线性垂直布局，向该布局中添加四个 ImageView 控件。

布局文件代码如下：

```xml
<?xml version="1.0" encoding="utf-8"?>
<LinearLayout
    xmlns:android="http://schemas.android.com/apk/res/android"
    xmlns:app="http://schemas.android.com/apk/res-auto"
    xmlns:tools="http://schemas.android.com/tools"
    android:layout_width="match_parent"
    android:layout_height="match_parent"
    android:orientation="vertical"
    tools:context=".MainActivity">
    <ImageView
        android:layout_width="200dp"
        android:layout_height="100dp"
        android:foreground="@drawable/a" />
    <ImageView
        android:layout_width="200dp"
        android:layout_height="200dp"
        android:background="@drawable/a" />
    <ImageView
        android:layout_width="200dp"
        android:layout_height="100dp"
        android:scaleType="fitCenter"
        android:src="@drawable/a" />
    <ImageView
        android:layout_width="200dp"
        android:layout_height="200dp"
        android:src="@drawable/a" />
</LinearLayout>
```

ImageView 示例运行结果如图 2.26 所示。

代码解析：图片"a.jpeg"的分辨率是 600×600（像素），程序中第一个控件将图片作为背景，所以进行了不等比拉伸；第二个控件图片的长宽比与图片相同，所以作为前景进行了等比拉伸；第三个控件尽管长宽比与图片不同，但因为设置了拉伸属性，所以图片进行了居中拉伸；第四个控件图片进行了等比拉伸。

图 2.26　ImageView 示例运行结果

2.3.5　RadioButton 控件

RadioButton（单选按钮）控件继承于 Button 类，可以直接使用 Button 支持的各种属性和方法。RadioButton 与普通按钮不同的是，它多了一个可以选中的功能，能额外指定一个 android:checked 属性。该属性可以指定初始状态是否被选中，默认的初始状态为不选中。

RadioButton 控件必须和单选框 RadioGroup 容器一起使用。在 RadioGroup 容器中可以放置多个 RadioButton 控件，但它们不可以同时被选中。

RadioGroup 容器继承于 LinearLayout，同样可以使用 orientation 属性设置内部 RadioButton 控件的排列方向。

RadioButton 控件的状态变化可以通过为 RadioGroup 容器设置监听事件 setOnChecked

ChangeListener()来处理。

【例 2.11】RadioButton 示例。

（1）界面设计。

创建应用程序，编辑 XML 布局文件，使用 LinearLayout 线性垂直布局，依次向该布局中添加一个 TextView 控件用于显示标题、一个 RadioGroup 容器用于容纳两个 RadioButton 控件和一个 TextView 控件显示选择结果。

布局文件代码如下：

```xml
<?xml version="1.0" encoding="utf-8"?>
<LinearLayout
    xmlns:android="http://schemas.android.com/apk/res/android"
    xmlns:app="http://schemas.android.com/apk/res-auto"
    xmlns:tools="http://schemas.android.com/tools"
    android:layout_width="match_parent"
    android:layout_width="match_parent"
    android:layout_height="match_parent"
    android:orientation="vertical"
    tools:context=".MainActivity">
    <TextView
        android:id="@+id/textView1"
        android:layout_width="wrap_content"
        android:layout_height="wrap_content"
        android:layout_marginTop="44dp"
        android:text="性别："
        android:textSize="20dp" />
    <RadioGroup
        android:id="@+id/radioGroup1"
        android:layout_width="wrap_content"
        android:layout_height="wrap_content"
        android:orientation="horizontal">
        <RadioButton
            android:id="@+id/radio0"
            android:layout_width="wrap_content"
            android:layout_height="wrap_content"
            android:checked="true"
            android:text="男" />
        <RadioButton
            android:id="@+id/radio1"
            android:layout_width="wrap_content"
            android:layout_height="wrap_content"
            android:text="女" />
    </RadioGroup>
    <TextView
        android:id="@+id/textView"
        android:layout_width="wrap_content"
        android:layout_height="wrap_content"
        android:text="您的性别是：男"
        android:textSize="30sp" />
</LinearLayout>
```

（2）设置监听器。

编辑 MainActivity 类代码，添加属性如下：

```
private TextView resultTextView;
private RadioGroup radioGroup;
```

修改 onCreate()方法，代码如下：

```
@Override
protected void onCreate(Bundle savedInstanceState) {

    super.onCreate(savedInstanceState);
    setContentView(R.layout.activity_radio_button);

    this.resultTextView = (TextView) findViewById(R.id.textView);
    this.radioGroup = (RadioGroup) findViewById(R.id.radioGroup1);
    this.radioGroup.setOnCheckedChangeListener(new
            RadioGroup.OnCheckedChangeListener() {
        @Override
        public void onCheckedChanged(RadioGroup radioGroup, int i) {
            if (i == R.id.radio0) {
                resultTextView.setText("您的性别是：男");
            } else {
                resultTextView.setText("您的性别是：女");
            }
        }
    });
}
```

RadioButton 示例运行结果如图 2.27 所示。

在上述代码中，使用 RadioGroup 的 setOnCheckedChangeListener()方法为其设置事件监听器，实现 RadioGroup.OnCheckedChangeListener 接口中的 onCheckedChanged 事件处理方法。此方法的第一个参数为事件源容器，第二个参数为事件源控件（RadioButton 控件）id，在事件处理方法内部编写代码，以实现当 RadioButton 控件选中状态改变时，可动态修改选中信息的逻辑代码。

图 2.27　RadioButton 示例运行结果

2.3.6　CheckBox 控件

CheckBox（复选框）控件继承于 Button 类，实现多选功能，每个 CheckBox 控件都有选中和未选中两种状态。在 XML 布局文件中可通过 CheckBox 控件的 checked 属性设定，当 checked 属性设置为 true 时，则表示选中；设置为 false 时，则表示未被选中。在程序中可通过 isChecked()方法来判断 CheckBox 控件是否被选中。

通常为 CheckBox 控件绑定 CompoundButton.OnCheckedChangeListener 监听器，当用户单击时可在两种状态间进行切换，触发 OnCheckedChange 事件，调用 onCheckedChanged(CompoundButton buttonView,boolean isChecked)方法，其中，第一个参数用于确定哪个 CheckBox 的状态发生了改变，第二个参数用于确定该 CheckBox 控件的状态值。

【例 2.12】CheckBox 示例。

（1）界面设计。

创建应用程序，编辑 XML 布局文件。使用 LinearLayout 线性垂直布局，向布局中添加两个 TextView 控件，一个用于显示标题，另一个用于显示选择结果。添加两个 CheckBox 控件，check 属性默认为 false。布局文件代码如下：

```xml
<?xml version="1.0" encoding="utf-8"?>
<LinearLayout xmlns:android="http://schemas.android.com/apk/res/android"
    xmlns:app="http://schemas.android.com/apk/res-auto"
    xmlns:tools="http://schemas.android.com/tools"
    android:layout_width="match_parent"
    android:layout_height="match_parent"
    android:orientation="vertical"
    tools:context=".MainActivity">
    <TextView
        android:layout_width="wrap_content"
        android:layout_height="wrap_content"
        android:text="您喜欢的水果是："
        android:textSize="40sp" />
    <CheckBox
        android:id="@+id/apple"
        android:layout_width="wrap_content"
        android:layout_height="wrap_content"
        android:text="苹果"
        android:textSize="30sp" />
    <CheckBox
        android:id="@+id/banana"
        android:layout_width="wrap_content"
        android:layout_height="wrap_content"
        android:text="香蕉"
        android:textSize="30sp" />
    <TextView
        android:id="@+id/resultTxtView"
        android:layout_width="wrap_content"
        android:layout_height="wrap_content"
        android:textSize="35sp" />
</LinearLayout>
```

（2）为 CheckBox 控件设置事件监听器。

编辑 MainActivity 类代码，添加私有属性如下：

```java
private TextView resultTextView; // 结果显示控件
private String hobbys = new String(); // 结果内容
```

设置 MainActivity 类实现 CompoundButton.OnCheckedChangeListener 接口，表示它同时也是 CheckBox 控件的监听器，在 onCreate()方法中初始化私有实例，同时为 CheckBox 控件设置事件监听器，代码如下：

```java
@Override
protected void onCreate(Bundle savedInstanceState) {
    super.onCreate(savedInstanceState);
    setContentView(R.layout.activity_check_button);
    resultTextView = (TextView) findViewById(R.id.resultTxtView);
    CheckBox appleCheckButton = (CheckBox) findViewById(R.id.apple);
```

```
            appleCheckButton.setOnCheckedChangeListener(this);
            CheckBox bananaCheckButton = (CheckBox) findViewById(R.id.banana);
            bananaCheckButton.setOnCheckedChangeListener(this);
        }
```

实现 OnCheckedChangeListener 接口中声明的事件处理方法，代码如下：

```
        @Override
        public void onCheckedChanged(CompoundButton compoundButton, boolean b) {
            String selectTxt = compoundButton.getText().toString();
            if (b) {
                if (!this.hobbys.contains(selectTxt)) {
                    this.hobbys += selectTxt;
                    this.resultTextView.setText(hobbys);
                }
            } else {
                if (this.hobbys.contains(selectTxt)) {
                    this.hobbys.replace(selectTxt, "");
                    this.resultTextView.setText(hobbys);
                }
            }
        }
```

在上述代码中，onCheckedChanged()方法的第一个参数表示事件源，第二个参数表示代码的选中状态。根据其值获取 CheckBox 控件的选中状态，通过动态更新显示结果控件中的内容。

CheckBox 示例运行结果如图 2.28 所示。

2.3.7　Toast 控件

Toast（消息提示）控件在 Android 中用于提示信息。在屏幕中显示一个消息提示框，没有任何按钮，也不会获得焦点，一段时间后便会自动消失，常常用于显示网络未连接、用户密码输入错误或退出应用等提示性信息。

普通的 Toast 对象可以通过如下静态方法创建：

```
        Toast.makeText(context,text,time)
```

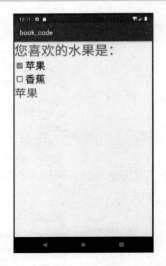

图 2.28　CheckBox 示例运行结果

context：表示应用程序上下文环境信息，在 Activity 类中可以使用"当前 Activity 类名称.this"来表示。

text：表示 Toast 控件中显示的文本信息。

time：表示 Toast 控件显示的时间长短，其取值可以是 Toast.LENGTH_SHORT 或 Toast.LENGTH_LONG，相对应的时间过去后 Toast 控件就会自动消失。

需要注意的是，取得 Toast 对象后 Toast 控件并未显示出来，必须手动调用 show()方法才可显示。

Toast 控件默认显示在屏幕底部，可以使用 setGravity()方法设置 Toast 控件的显示位置，如开发人员可以通过执行代码"toast.setGravity(Gravity.CENTER, 0, 0);"使 Toast 控件显示在屏幕中部。

除了基本 Toast 控件，还可以设置其内部显示信息的格式。Toast 控件甚至可以加载用户自定义的 XML 布局文件，下面的示例展示了基本 Toast 控件和用户自定义 Toast 控件的基本操作。

【例 2.13】Toast 示例。

（1）创建应用程序，编辑主界面布局文件。

在 Android Studio 中创建一个应用程序，编辑 activity_main.xml 布局文件。使用 LinearLayout 在其中添加两个 Button 按钮，通过单击可分别显示两种 Toast 控件。

布局文件的代码如下：

```xml
<?xml version="1.0" encoding="utf-8"?>
<LinearLayout
    xmlns:android="http://schemas.android.com/apk/res/android"
    xmlns:app="http://schemas.android.com/apk/res-auto"
    xmlns:tools="http://schemas.android.com/tools"
    android:layout_width="match_parent"
    android:layout_height="match_parent"
    android:orientation="vertical"
    android:padding="20dp"
    tools:context=".MainActivity">
    <Button
        android:id="@+id/toastBtn1"
        android:layout_width="wrap_content"
        android:layout_height="wrap_content"
        android:textSize="25sp"
        android:text="基本 Toast" />
    <Button
        android:id="@+id/toastBtn2"
        android:layout_width="wrap_content"
        android:layout_height="wrap_content"
        android:textSize="25sp"
        android:text="自定义 Toast" />
</LinearLayout>
```

（2）自定义布局文件。

选择 res/layout 目录，右击，在弹出的菜单中执行"new"→"XML"→"Layout XML File"菜单命令，如图 2.29 所示。

在对话框中输入布局文件名 toast_suxtomer，单击"Finish"按钮，创建一个新的布局文件。该布局文件使用 LinearLayout，在其中增加一个 ImageView 控件显示 Toast 图标和一个 TextView 控件显示 Toast 的提示信息。布局文件 toast_customer.xml 具体代码如下：

```xml
<?xml version="1.0" encoding="utf-8"?>
<LinearLayout xmlns:android="http://schemas.android.com/apk/res/android"
    android:layout_width="match_parent"
    android:layout_height="match_parent"
    android:id="@+id/toast_customer">
    <ImageView
        android:id="@+id/img"
        android:layout_width="24dp"
        android:layout_height="24dp"
        android:layout_marginLeft="10dp" />
    <TextView
        android:id="@+id/tv_msg"
        android:layout_width="match_parent"
        android:layout_height="wrap_content"
```

```
            android:layout_marginLeft="10dp"
            android:textSize="20sp" />
</LinearLayout>
```

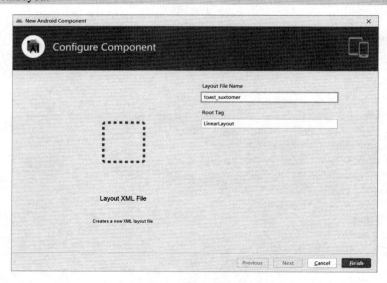

图 2.29 新建布局文件

(3) 编辑 MainActivity。

MainActivity 类实现 View.OnClickListener 接口,可作为按钮单击事件的监听器,实现事件处理方法 onClick()。MainActivity 代码如下:

```
public class MainActivity extends AppCompatActivity implements View.OnClickListener {
    @Override
    protected void onCreate(Bundle savedInstanceState) {
        super.onCreate(savedInstanceState);
        setContentView(R.layout.activity_main);
        Button normalToastBtn = (Button) findViewById(R.id.toastBtn1);
        Button customToastBtn = (Button) findViewById(R.id.toastBtn2);
        normalToastBtn.setOnClickListener(this);
        customToastBtn.setOnClickListener(this);
    }
    @Override
    public void onClick(View v) {
        public void onClick(View v) {
        switch (v.getId() ) {
            case R.id.toastBtn1:
                Toast.makeText(MainActivity.this,
                    "普通 Toast 控件", Toast.LENGTH_LONG).show();
                break;
            case R.id.toastBtn2:
                LayoutInflater inflater = getLayoutInflater();
                View toastView = inflater.inflate(R.layout.toast_customer,null);
                ImageView img = (ImageView) toastView.findViewById(R.id.img);
                img.setImageResource(R.drawable.qq);
                TextView tv_msg = (TextView) toastView.findViewById(R.id.tv_msg);
```

```
                    tv_msg.setText("用户自定义Toast");
                    Toast toast = new Toast(MainActivity.this);
                    toast.setGravity(Gravity.CENTER, 0, 0);
                    toast.setDuration(Toast.LENGTH_LONG);
                    toast.setView(toastView);
                    toast.show();
                      break;
            }
        }
    }
```

上述代码的 onClick()方法中编写事件处理逻辑代码。在第二个按钮的事件处理代码中，加载用户自定义的 toast_customer.xml 布局，并为布局中的 ImageView 控件设置图片资源（需要先将 qq.jpeg 复制到 res/drawable 目录中）和 TextView 控件设置文字内容，然后使用 new 运算符构造出 Toast 控件类对象，参数是程序运行上下文环境，通过 Toast 控件的成员方法设置显示位置、持续显示时间和设置视图，最后调用 show 方法将 Toast 控件显示出来。

运行程序，单击主界面的"基本 Toast"按钮后，在屏幕下部正中位置显示一个普通 Toast 控件，效果如图 2.30 所示。单击"自定义 Toast"按钮，在屏幕中部显示一个用户自定义的 Toast 控件，效果如图 2.31 所示。

图 2.30　普通 Toast 控件

图 2.31　用户自定义 Toast

习题 2

1．设计一个考试系统登录界面。

（1）界面包括应用项目的名称，"账号"和"密码"输入编辑框，"登录"和"注册"按钮，"记住密码"文本框，"是"和"否"单选按钮，如图 2.32 所示。"账号"和"密码"输入编辑框内有提示信息。为了界面美观，可以配上合适的背景图。

（2）编写"登录"按钮的单击事件代码。如果账号或密码没有填写，则用 Toast 控件显示提示信息"账号和密码不能为空"。如果账号或密码错误，则显示提示信息"账号或密码错误！"，如果填写正确，则显示"登录成功！"，如图 2.33 所示。

图 2.32　登录界面　　　　　　　　　图 2.33　填写信息

2．设计一个爱好调查界面。

（1）界面包括一个用于输入姓名的编辑框，一组用于选择性别的单选按钮，一组用于选择爱好的复选框，一个"确定"按钮，下部有一个输入编辑框用于显示信息，如图 2.34 所示。

（2）在编辑框内填写姓名，在单选按钮组中选择性别，在复选框中选择爱好，单击"确定"按钮，结果显示在下部的编辑框内，如图 2.35 所示。

3．设计计算器界面，如图 2.36 所示。

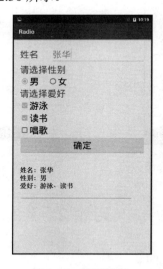

图 2.34　爱好调查界面　　　　图 2.35　填写信息　　　　图 2.36　计算器界面

第 3 章　用户界面高级控件

随着界面功能复杂度的提高，仅使用基本控件已很难满足设计的需要，同时手持智能终端因其显示尺寸受限往往对控件有着特殊的要求。因此，Android SDK 为设计人员提供了一组高级用户界面控件，另外，开发人员还可以开发自定义控件。本章主要介绍对话框、状态栏通知、日期/时间选择器、滚动条、进度条、列表视图等常用高级控件和自定义控件。

3.1 弹出式控件

通常移动智能终端的显示面积都十分有限，因此可以使用弹出式复合控件来容纳更多的显示内容。

3.1.1 AlertDialog 控件

AlertDialog（对话框）控件在 Windows 程序设计中被大量应用，它在 Android 中的对话框常用于显示提示信息或其他一些重要的用户交互内容，一般以小窗口的形式显示在界面窗口之上。

Android SDK 提供 AlertDialog 类来完成对话框的设计。普通对话框由标题栏、内容区和按钮区组成，在标题栏中可以显示图标和标题名称,内容区中显示提示信息或放置单选框或多选框等基本用户组件以完成交互操作。按钮区可以根据需要设置"确定"或"取消"按钮。Android 对话框的组成如图 3.1 所示。

AlertDialog 类的构造方法不是公有的，Android 创建对话框对象采用了工厂类设计模式，需要通过 AlertDialog.Builder 类来完成对话框对象的创建和初始化操作，具体步骤如下。

图 3.1　Android 对话框的组成

（1）获取 AlertDialog.Builder 对象。
（2）使用 Builder 对象的 setTitle()方法设置标题，并使用 setIcon()方法设置图标完成标题栏的设计。
（3）使用 Builder 对象的 setMessage()方法设置提示内容,或者使用 setSingleChoiceItems()方法和 setMultiChoiceItems()方法设置对话框的内容区。
（4）使用 Builder 对象的 setPositiveButton()方法和 setNegativeButton()方法设置按钮区内的按钮。
（5）调用 create()方法得到 AlertDialog 控件。

(6) 调用 AlertDialog 控件的 show()方法来显示对话框。

不仅如此,当 AlertDialog 控件提供的功能无法满足业务逻辑的需要时,开发人员可以继承 AlertDialog 类实现用户自定义对话框,下面通过示例来说明各种对话框的使用。

【例 3.1】AlertDialog 控件的示例。

(1) 创建应用程序,编辑主界面布局。

创建一个应用程序,编辑 activity_main.xml 布局文件,使用 RelativeLayout 布局容器,在其中添加四个 Button 控件,分别单击显示普通对话框、单选对话框、复选对话框和用户自定义对话框。activity_main.xml 布局文件代码如下:

```xml
<?xml version="1.0" encoding="utf-8"?>
<RelativeLayout xmlns:android="http://schemas.android.com/apk/res/android"
    xmlns:app="http://schemas.android.com/apk/res-auto"
    xmlns:tools="http://schemas.android.com/tools"
    android:layout_width="match_parent"
    android:layout_height="match_parent"
    tools:context=".MainActivity">
    <Button
        android:id="@+id/normalDlgBtn"
        android:layout_width="wrap_content"
        android:layout_height="wrap_content"
        android:layout_centerHorizontal="true"
        android:layout_marginTop="10dp"
        android:textSize="20sp"
        android:text="普通对话框" />
    <Button
        android:id="@+id/singleDlgBtn"
        android:layout_width="wrap_content"
        android:layout_height="wrap_content"
        android:layout_below="@id/normalDlgBtn"
        android:layout_alignLeft="@id/normalDlgBtn"
        android:layout_marginTop="10dp"
        android:textSize="20sp"
        android:text="单选对话框" />
    <Button
        android:id="@+id/multiplyDlgBtn"
        android:layout_width="wrap_content"
        android:layout_height="wrap_content"
        android:layout_below="@id/singleDlgBtn"
        android:layout_alignLeft="@id/singleDlgBtn"
        android:layout_marginTop="10dp"
        android:textSize="20sp"
        android:text="多选对话框" />
    <Button
        android:id="@+id/customDlgBtn"
        android:layout_width="wrap_content"
        android:layout_height="wrap_content"
        android:layout_below="@id/multiplyDlgBtn"
        android:layout_alignLeft="@id/multiplyDlgBtn"
        android:layout_marginTop="10dp"
        android:textSize="20sp"
        android:text="用户自定义对话框" />
</RelativeLayout>
```

（2）编辑 MainActivity 类代码。

MainActivity 类实现 View.OnClickListener 接口，作为按钮单击事件监听器。在 MainActivity 类中声明以下变量：

```java
String sex[]={"男","女"};
int sexIndex = 0;
String sports[]={"跑步","跳远","足球","游泳"};
boolean selected[]={false,false,false,false};
```

在 onCreate()方法内部为四个按钮添加单击事件监听器，代码如下：

```java
Button normalDlgBtn = (Button) findViewById(R.id.normalDlgBtn);
Button singleDlgBtn = (Button) findViewById(R.id.singleDlgBtn);
Button multiDlgBtn = (Button) findViewById(R.id.multiplyDlgBtn);
Button customerDlgBtn = (Button) findViewById(R.id.customDlgBtn);
normalDlgBtn.setOnClickListener(this);
singleDlgBtn.setOnClickListener(this);
multiDlgBtn.setOnClickListener(this);
customerDlgBtn.setOnClickListener(this);
```

（3）编写 onClick()方法。

在 onClick()方法中分别为四个按钮添加事件处理逻辑代码。

① "普通对话框" 按钮逻辑代码。

```java
if (v.getId() == R.id.normalDlgBtn) {
    AlertDialog.Builder builder = new    AlertDialog.Builder(MainActivity.this);
    builder.setTitle("普通对话框")
            .setIcon(R.mipmap.ic_launcher)
            .setMessage("是否退出程序？")
            .setPositiveButton("确定", new DialogInterface.OnClickListener() {
                @Override
                public void onClick(DialogInterface dialog,
                                    int which) {
                    Toast.makeText(MainActivity.this,
                            "您单击了普通对话框的确认按钮",
                            Toast.LENGTH_LONG).show();
                }
            })
            .setNegativeButton("取消", new DialogInterface.OnClickListener() {
                @Override
                public void onClick(DialogInterface dialog,
                                    int which) {
                    Toast.makeText(MainActivity.this,
                            "您单击了普通对话框的取消按钮",
                            Toast.LENGTH_LONG).show();
                }
            })
            .show();
}
```

在 Builder 类添加按钮的方法中，第一个参数代表按钮上的显示文字，第二个参数是按钮事件监听器，是一个实现 DialogInterface.OnClickListener 接口的类，开发人员需要按照程序逻辑的需要重写按钮单击事件处理方法 onClick()。

运行程序单击第一个按钮的显示结果如图 3.2 所示。

②"单选对话框"按钮逻辑代码。

```
if (v.getId() == R.id.singleDlgBtn) {
    AlertDialog.Builder builder = new AlertDialog.Builder(MainActivity.this);
    builder.setIcon(R.mipmap.ic_launcher)
        .setTitle("性别")
        .setSingleChoiceItems(sex, 0, new DialogInterface.OnClickListener() {
            @Override
            public void onClick(DialogInterface dialogInterface, int which) {
                sexIndex = which;
            }
        })
        .setPositiveButton("确定", new DialogInterface.OnClickListener() {
            @Override
            public void onClick(DialogInterface dialogInterface, int which) {
                Toast.makeText(MainActivity.this,
                        "您的性别是:  " + sex[which],
                        Toast.LENGTH_LONG).show();
            }
        })
        .setNegativeButton("取消", new DialogInterface.OnClickListener() {
            @Override
            public void onClick(DialogInterface dialogInterface, int which) {
                Toast.makeText(MainActivity.this,
                        "您单击了单选对话框的取消按钮",
                        Toast.LENGTH_LONG).show();
            }
        })
        .show();
}
```

在这里使用了 Builder 对象的 setSingleChoiceItems()方法来设置内容区域内的单选控件，其中，第一个参数代表每个单选按钮的标题，是一个字符串数组，第二个参数代表了默认选中项，第三个参数是单选按钮单击监听器，是一个实现了 DialogInterface.OnClickListener 接口的对象，开发人员需要重写事件处理方法 onClick()，此方法中第一个参数是单选按钮对象，第二个参数是被单击的单选按钮的索引，用户可以据此来判断选中项，运行程序结果如图 3.3 所示。

图 3.2　普通对话框示例

图 3.3　单选对话框示例

③ "多选对话框"按钮逻辑代码。

```java
if (v.getId() == R.id.multiplyDlgBtn) {
    AlertDialog.Builder builder = new AlertDialog.Builder(MainActivity.this);
    builder.setIcon(R.mipmap.ic_launcher)
            .setTitle("运动项目")
            .setMultiChoiceItems(sports, selected,
                    new DialogInterface.OnMultiChoiceClickListener() {
                        @Override
                        public void onClick(DialogInterface dialog, int which, boolean isChecked) {
                            selected[which]=isChecked;
                        }
                    })
            .setPositiveButton("确定",
                    new DialogInterface.OnClickListener() {
                        @Override
                        public void onClick(DialogInterface dialog, int which) {
                            StringBuilder sb = new StringBuilder();
                            for (int ind = 0; ind < selected.length; ++ind) {
                                if (selected[ind]) {
                                    sb.append(sports[ind]).append(" ");
                                }
                            }
                            Toast.makeText(MainActivity.this,
                                    sb.toString(),
                                    Toast.LENGTH_LONG).show();
                        }
                    })
            .setNegativeButton("取消",
                    new DialogInterface.OnClickListener() {
                        @Override
                        public void onClick(DialogInterface dialog, int which) {
                            Toast.makeText(MainActivity.this,
                                    "您单击了取消按钮",
                                    Toast.LENGTH_LONG).show();
                        }
                    })
            .show();
}
```

在多选对话框中使用 Builder 类的 setMultiChoiceItems() 方法设置多选控件。该方法的第一个参数表示多选控件的显示内容，是一个字符串数组；第二个参数表示默认选中项，是一个 Boolean 类型的数组，这两个参数数组长度相同；第三个参数是一个多选控件单击监听器，它是一个实现了 DialogInterface.OnMultiChoiceClickListener 接口的对象。开发人员需要重写事件处理方法 onClick()，此方法有三个参数：第一个参数表示事件源，第二个参数代表多选按钮的索引，第三个参数表示其选中状态，true 表示选中，false 表示未选中。

运行程序，单击"多选对话框"按钮弹出多选对话框，效果如图 3.4 所示。

图 3.4　多选对话框

④ "用户自定义对话框"按钮的逻辑代码。

用户自定义对话框可以根据程序需要使用所有的用户控件，与 Activity 一样，对话框也可以使用 XML 布局文件设计界面。选择 res/layout 目录，右击，执行"new"→"XML"→"Layout XML File"菜单命令，创建布局文件 customer_dialog.xml，其代码如下：

```xml
<?xml version="1.0" encoding="utf-8"?>
<LinearLayout xmlns:android="http://schemas.android.com/apk/res/android"
    android:layout_width="match_parent"
    android:layout_height="match_parent"
    android:orientation="vertical">
    <LinearLayout
        android:layout_width="match_parent"
        android:layout_height="wrap_content"
        android:orientation="vertical"
        android:paddingTop="20dp">
        <TextView
            android:id="@+id/title"
            android:layout_width="match_parent"
            android:layout_height="wrap_content"
            android:layout_marginBottom="15dp"
            android:gravity="center"
            android:textSize="25sp" />
        <TextView
            android:id="@+id/message"
            android:layout_width="match_parent"
            android:layout_height="wrap_content"
            android:layout_marginLeft="15dp"
            android:layout_marginRight="15dp"
            android:textColor="#CCCCCC"
            android:textSize="15sp" />
        <View
            android:layout_width="match_parent"
            android:layout_height="2px"
            android:layout_marginTop="10dp"
```

```xml
                android:background="#E8E8E8" />
            <LinearLayout
                android:layout_width="match_parent"
                android:layout_height="wrap_content">
                <Button
                    android:id="@+id/cancel"
                    android:layout_width="0dp"
                    android:layout_height="wrap_content"
                    android:layout_marginTop="10dp"
                    android:layout_marginLeft="10dp"
                    android:layout_marginBottom="15dp"
                    android:layout_weight="1"
                    android:background="@null"
                    android:gravity="center"
                    android:textSize="15sp" />
                <Button
                    android:id="@+id/ok"
                    android:layout_width="0dp"
                    android:layout_height="wrap_content"
                    android:layout_marginTop="10dp"
                    android:layout_marginRight="10dp"
                    android:layout_marginBottom="15dp"
                    android:layout_weight="1"
                    android:background="@null"
                    android:gravity="center"
                    android:textSize="15sp" />
            </LinearLayout>
        </LinearLayout>
    </LinearLayout>
```

界面中包括两个 TextView 控件用于显示标题和提示内容,这两个按钮控件(确认和取消)之间使用 View 作为分割线。布局文件在对话框类中挂载即可显示用户设计的对话框界面。

创建 CustomerDialog 类,指定其父类为 AlertDialog,添加私有属性表示界面中的四个控件和它们显示文本的内容,同时为这些私有属性添加 setter 方法,其代码如下:

```java
    private TextView title, message;
    private Button ok, cancel;
    private String strTitle, strMessage, strOk, strCancel;
    public void setTitle(String title) {
            strTitle = title;
    }
    public void setMessage(String message) {
            strMessage = message;
    }
    public void setOk(String ok) {
            strOk = ok;
    }
    public void setCancel(String cancel) {
            strCancel = cancel;
    }
```

在 CustomerDialog 类的 onCreate()方法中对这些对象进行初始化,其代码如下:

```java
    @Override
    protected void onCreate(Bundle savedInstanceState) {
```

```
        super.onCreate(savedInstanceState);
        setContentView(R.layout.customer_dialog);
        title = (TextView) findViewById(R.id.title);
        message = (TextView) findViewById(R.id.message);
        ok = (Button) findViewById(R.id.ok);
        cancel = (Button) findViewById(R.id.cancel);
    }
```

编写 show()方法，为控件设置显示内容，代码如下：

```
    @Override
    public void show() {
        super.show();
        title.setText(strTitle);
        message.setText(strMessage);
        ok.setText(strOk);
        cancel.setText(strCancel);
    }
```

此时，在 MainActivity 中添加显示用户自定义对话框的逻辑代码，代码如下：

```
    if (view.getId() == R.id.customDlgBtn) {
        CustomerDialog dlg = new CustomerDialog(MainActivity.this);
        dlg.setTitle("标题");
        dlg.setMessage("提示信息...");
        dlg.setOk("确认");
        dlg.setCancel("取消");
        dlg.show();
    }
```

运行程序，单击"用户自定义对话框"按钮，显示结果如图 3.5 所示。

图 3.5　用户自定义对话框

3.1.2　Notification 控件

Android 系统使用状态栏通知显示信息，效果如图 3.6 所示。

Android 中的 Notification（状态栏通知）控件分为三种，包括普通 Notification、折叠

Notification 和悬浮 Notification。在图中可以看出尽管 Notification 控件显示面积很小，但是其中包含了很多种子控件，如标题、内容文本、时间戳、大图标、小图标和应用名称等。这些子控件都可以通过 Notification 类的成员方法来设置，如同 AlertDialog 类一样 Notification 控件也是通过工厂类来获取的，创建 Notification 控件的代码如下：

```
new NotificationCompat.Builder(this, "channelId").build();
```

Builder 类构造方法的第一个参数是上下文环境，第二个参数是通知渠道，从 Android 8.0（API 级别 26）开始，必须为所有通知分配渠道（NotificationChannel），否则通知将不会显示。

Android 利用通知的重要程度来决定在多大程度上干扰用户（视觉上和听觉上）。通知的重要程度越高，干扰程度就越高。在 Android 8.0（API 级别 26）及更高版本的设备上，通知的重要程度由发布渠道 importance 决定，用户可以在系统设置中更改通知渠道的重要程度。在 Android 7.1（API 级别 25）及更低版本的设备上，每条通知的重要程度均由通知的 priority 决定。

图 3.6 状态栏通知的显示效果

创建 Notification 控件需要先使用 NotificationCompat.Builder 对象设置通知内容和渠道，然后通过以下方法设置通知的各个元素。

小图标：通过 setSmallIcon()设置。这是用户唯一的可见内容。

标题：通过 setContentTitle()设置。

正文文本：通过 setContentText()设置。

通知优先级：通过 setPriority()设置。优先级可以确定通知在 Android 7.1 和更低版本上的干扰程度（对于 Android 8.0 和更高版本，必须设置渠道重要性）。

显示 Notification 控件需要使用 NotificationManager 对象，通过调用 Activity 类的 getSystemService(Context.NOTIFICATION_SERVICE)获取一个 NotificationManager 对象，调用 NotificationManager 对象的 createNotificationChannel()方法注册该通知渠道。通知渠道注册后就无法修改了，再调用 NotificationManagerCompat.notify()，并将通知的唯一 ID（自定义的一个整数）和 NotificationCompat.Builder.build() 的结果传递给它用于显示一条 Notification。

【例 3.2】Notification 控件的示例。

首先，获取 NotificationManager 对象，代码如下：

```
NotificationManager manager =
    (NotificationManager) getSystemService(NOTIFICATION_SERVICE);
```

然后判断 Android 平台版本，创建通知渠道，代码如下：

```
if (Build.VERSION.SDK_INT >= Build.VERSION_CODES.O) {
    NotificationChannel channel = new NotificationChannel("channelId",
            "一个通知",
            NotificationManager.IMPORTANCE_HIGH);
    manager.createNotificationChannel(channel);
}
```

最后，通过 Builder 类创建 Notification 控件，并使用 NotificationManager 对象显示出来。

完整代码如下：
```java
public class MainActivity extends AppCompatActivity {
    @Override
    protected void onCreate(Bundle savedInstanceState) {
        super.onCreate(savedInstanceState);
        setContentView(R.layout.activity_main);
        NotificationManager manager =
          (NotificationManager) getSystemService(NOTIFICATION_SERVICE);
        if (Build.VERSION.SDK_INT >= Build.VERSION_CODES.O) {
            NotificationChannel channel = new NotificationChannel("channelId",
                "一个通知",
                NotificationManager.IMPORTANCE_HIGH);
            manager.createNotificationChannel(channel);
        }
        Notification notification = new NotificationCompat.Builder(this, "channelId")
                .setContentTitle("通知")
                .setContentText("赶紧来上课")
                .setSmallIcon(R.mipmap.ic_launcher)
                .setLargeIcon(BitmapFactory.decodeResource(getResources(),
                        R.drawable.qq))
                .build();
        manager.notify(1, notification);
    }
}
```

运行程序，显示结果如图 3.7 所示。

图 3.7 Notification 控件的示例

3.2 日期/时间选择器

在输入信息时经常要输入日期/时间信息，而由于文化背景和习惯的不同，用户在书写日期/时间时格式往往不同，这时程序会误认为是用户输入的不同值，而无法处理用户输入的信息。因此，通常不允许用户直接输入日期/时间信息，而是使用日期/时间选择器进行选择。

3.2.1 DatePicker 控件

Android 提供了 DatePicker（日期选择）控件用于用户选择日期。在 XML 布局文件中添加 DataPicker 控件的代码如下：

```
<DatePicker
    android:id="@+id/datePicker"
    android:layout_width="match_parent"
    android:layout_height="match_parent" />
```

DatePicker 控件的常用属性如表 3.1 所示。

表 3.1 DatePicker 控件的常用属性

属　　性	说　　明
android:datePickerMode	外观样式，可选值包括 spinner 和 calendar（默认）
android:calendarViewShown	是否显示日历视图
android:spinnersShown	是否显示 spinner
android:calendarTextColor	日历列表的文本颜色
android:dayOfWeekBackground	顶部星期几的背景颜色
android:dayOfWeekTextAppearance	顶部星期几的文字颜色
android:firstDayOfWeek	日历列表以星期几开头
android:headerBackground	整个头部的背景颜色
android:headerDayOfMonthTextAppearance	头部日期字体的颜色
android:headerMonthTextAppearance	头部月份的字体颜色
android:headerYearTextAppearance	头部年份的字体颜色
android:maxDate	最大日期，mm / dd / yyyy 格式
android:minDate	最小日期，mm / dd / yyyy 格式
android:startYear	设置起始年，如 1994 年
android:yearListItemTextAppearance	年列表文本出现在列表中
android:yearListSelectorColor	年列表选择的颜色

如果要获取 DatePicker 控件选择的日期值需要使用监听器，该监听器实现了 DatePicker.OnDateChangedListener 接口，开发人员仅需要重写事件处理方法即可，事件处理方法的声明如下：

```
public void onDateChanged(DatePicker datePicker, int year, int month, int day)
```

其中参数 datePicker 表示事件源，参数 year、month 和 day 分别表示选择日期的年、月、日值。

【例 3.3】DatePicker 控件的示例。

新建一个 Activity，在 XML 布局文件中添加 DatePicker 控件，编写 Activity 类的 onCreate() 方法，代码如下：

```
public class MainActivity extends AppCompatActivity {
    @Override
    protected void onCreate(Bundle savedInstanceState) {
        super.onCreate(savedInstanceState);
        setContentView(R.layout.activity_main);
        DatePicker datePicker = (DatePicker) findViewById(R.id.datepicker);
        Calendar calendar = Calendar.getInstance();
```

```
                int year = calendar.get(Calendar.YEAR);
                int monthOfYear = calendar.get(Calendar.MONTH);
                int dayOfMonth = calendar.get(Calendar.DAY_OF_MONTH);
                datePicker.init(year,monthOfYear, dayOfMonth,
                        new DatePicker.OnDateChangedListener() {
                            @Override
                            public void onDateChanged(DatePicker datePicker,
                                    int year, int month, int day){
                                Toast.makeText(MainActivity.this,
                                    "您选择的日期是： " + year + "年"
                                    + (month + 1) + "月" + day + "日!",
                                    Toast.LENGTH_SHORT).show();
                            }
                        });
            }
        }
```

运行程序，显示结果如图 3.8 所示。

图 3.8　DatePicker 控件的示例

3.2.2　TimePicker 控件

TimePicker 控件（时间选择）可提供时间选择功能，在 XML 布局文件中的声明方式如下：

```
<TimePicker
    android:id="@+id/timePicker"
    android:layout_width="match_parent"
    android:layout_height="wrap_content"
    android:background="#ffffff"/>
```

TimePicker 控件自带两种风格，使用"android:timePickerMode"属性设置，其值为 clock 或 spinner，默认为 clock。

【例 3.4】TimePicker 控件的示例。

新建一个 Activity，在 XML 布局文件中添加 TimePicker 控件，编写 Activity 类的 onCreate() 方法，代码如下：

```java
public class MainActivity extends AppCompatActivity {
    @Override
    protected void onCreate(Bundle savedInstanceState) {
        super.onCreate(savedInstanceState);
        setContentView(R.layout.activity_main);
        TimePicker mTimepicker = (TimePicker)findViewById(R.id.timePicker);
        mTimepicker.setOnTimeChangedListener(
                new TimePicker.OnTimeChangedListener() {
                    @Override
                    public void onTimeChanged(TimePicker view,
                                              int hourOfDay, int minute){
                        Toast.makeText(MainActivity.this
                                ,"您选择的时间是:"+hourOfDay+"时"+minute+"分" ,
                                Toast.LENGTH_SHORT).show();
                    }
                });
    }
}
```

运行程序，显示结果如图 3.9 所示。

图 3.9　TimePicker 控件的示例

从上面的示例可以看出 Android 提供的 DatePicker 控件和 TimePicker 控件的界面效果都比较简单，仅适用于要求不高的场合。Android 还提供了 DatePickerDialog（日期对话框控件）和 TimePickerDialog（时间对话框控件）使开发人员有了更多选择。

3.3　滚动条和进度条

Android 程序一般都运行在显示受限的移动智能终端设备上，当屏幕无法容纳所需的程序界面时可以使用滚动条。同时，进度条可以在运行一个耗时操作时显示进度，否则会使用户产生程序卡顿的错觉。

3.3.1 ScrollView 控件

当绘制的 UI 控件超出手机屏幕尺寸时,就会导致 UI 控件无法显示。为了解决这个问题,Android 提供了 ScrollView(滚动条)控件,可以采用滑动的方式,使控件得以显示。

ScrollView 控件本质上是一个 FrameLayout 容器,只是在它的基础上添加了滚动,允许显示比实际多的内容。当 ScrollView 控件的内容大于其本身的尺寸时,就会自动添加滚动条,并可以竖直滑动。

ScrollView 控件的直接子 View(控件)只能有一个。也就是说,如果你要使用很复杂的视图结构,必须把这些视图放在一个标准布局里。可以使用 layout_width 和 layout_height 给 ScrollView 控件指定大小,ScrollView 控件只用来处理需要滚动的不规则视图的组合。大批量的列表数据展示可以使用 ListView、GridView 或 RecyclerView 实现。ScrollView 控件只支持垂直滑动,水平滑动使用 HorizontalScrollView 控件。

ScrollView 控件的 android:fillViewport 属性定义了是否可以拉伸其内容来填满 viewport,可以调用方法 setFillViewport(boolean)来达到一样的效果。

1. ScrollView 控件的常用属性

(1) android:fadingEdge="none"

设置拉滚动条时边框渐变的方向,其中 none 表示边框颜色不变,horizontal 表示水平方向颜色变淡,vertical 表示垂直方向颜色变淡。

(2) android:overScrollMode="never"

删除 ScrollView 控件拉到尽头(顶部、底部)后,继续拉时出现的阴影效果,适用于 Andriod 2.3 及以上版本,否则不用设置。

(3) android:scrollbars="none"

设置滚动条显示,其中 none 表示隐藏,horizontal 表示水平,vertical 表示垂直。

(4) android:descendantFocusability=""

当为 View 获取焦点时,可定义 ViewGroup 和其子控件两者之间的关系,其属性的值有以下三种。

① beforeDescendants:ViewGroup 会优先其子类控件而获取焦点;
② afterDescendants:ViewGroup 只有当其子类控件不需要获取焦点时才获取焦点;
③ blocksDescendants:ViewGroup 会覆盖子类控件而直接获得焦点。

(5) android:fillViewport="true"

这是 ScrollView 控件独有的属性,用来定义 ScrollView 控件是否需要拉伸来填充内容。

(6) android:fillViewport

是否允许 ScrollView 控件填充整个屏幕,如 ScrollView 控件嵌套的子控件高度达不到屏幕高度时,虽然 ScrollView 控件的高度设置了 match_parent,但也无法充满整个屏幕,只有设置 android:fillViewport="true"才能使 ScrollView 填满整个页面。

2. ScrollView 控件的常用方法

(1) fullScroll()

```
scrollView.fullScroll(ScrollView.FOCUS_DOWN);    //滚动到底部
scrollView.fullScroll(ScrollView.FOCUS_UP);      //滚动到顶部
```

(2) smoothScrollTo()

```
scrollView.post(new Runnable() {
```

```java
        @Override
        public void run() {
            //偏移值
            int offset = 100;
            scrollView.smoothScrollTo(0, offset);
        }});
```

（3）scrollView()

```java
scrollView.setOnTouchListener(new View.OnTouchListener() {
    @Override
    public boolean onTouch(View view, MotionEvent motionEvent) {
        // true 表示禁止滑动，false 表示可滑动
        return true;
    }
});
```

【例 3.5】ScrollView 控件的示例。

（1）创建 Activity，编辑 XML 布局文件代码。

创建一个应用程序，主界面使用垂直 LinearLayout 布局。首先添加两个按钮，单击相应按钮使 ScrollView 控件分别滚动到顶部和底部；然后添加 ScrollView 控件，并在其中添加一个 TextView 控件，在程序代码中添加 TextView 控件的内容，XML 布局代码如下：

```xml
<?xml version="1.0" encoding="utf-8"?>
<LinearLayout xmlns:android="http://schemas.android.com/apk/res/android"
    xmlns:app="http://schemas.android.com/apk/res-auto"
    xmlns:tools="http://schemas.android.com/tools"
    android:layout_width="match_parent"
    android:layout_height="match_parent"
    android:orientation="vertical"
    tools:context=".chap03.s2.ScrollViewActivity">
    <Button
        android:id="@+id/bottom"
        android:layout_width="match_parent"
        android:layout_height="wrap_content"
        android:text="底部" />
    <Button
        android:id="@+id/top"
        android:layout_width="match_parent"
        android:layout_height="wrap_content"
        android:text="顶部" />
    <ScrollView
        android:id="@+id/scrollView"
        android:layout_width="match_parent"
        android:layout_height="wrap_content">
        <TextView
            android:id="@+id/textView"
            android:layout_width="match_parent"
            android:layout_height="wrap_content"
            android:textSize="15sp" />
    </ScrollView>
</LinearLayout>
```

（2）编写 Activity 类。

在 onCreate()方法内先通过 findViewById 方法初始化按钮控件、ScrollView 控件和

TextView 控件,再设置 TextView 控件中显示的文本(此文本需要足够大)。然后为按钮控件设置事件监听器,在事件处理方法中使用 ScrollView 控件的 fullScroll 方法,将 ScrollView 控件滚动到相应的位置,代码如下:

```java
@Override
protected void onCreate(Bundle savedInstanceState) {
    super.onCreate(savedInstanceState);
    setContentView(R.layout.activity_scroll_view);
    StringBuilder sb = new StringBuilder();
    for (int i = 0; i < 100; ++i) {
        sb.append("Item" + i + "\n");
    }
    TextView textView = (TextView) findViewById(R.id.textView);
    textView.setText(sb.toString());
    Button bottomBtn = (Button) findViewById(R.id.bottom);
    Button topBtn = (Button) findViewById(R.id.top);
    ScrollView scrollView = (ScrollView) findViewById(R.id.scrollView);
    bottomBtn.setOnClickListener(new View.OnClickListener() {
        @Override
        public void onClick(View view) {
            scrollView.fullScroll(ScrollView.FOCUS_DOWN);
        }
    });
    topBtn.setOnClickListener(new View.OnClickListener() {
        @Override
        public void onClick(View view) {
            scrollView.fullScroll(ScrollView.FOCUS_UP);
        }
    });
}
```

运行程序,通过在屏幕上进行拖曳操作可以看到 ScrollView 控件出现了垂直滚动条,单击按钮 TextView 控件的内容时,可分别滚动到顶部和底部,程序运行效果如图 3.10 所示。

图 3.10 ScrollView 控件的示例

3.3.2 ProgressBar 控件

手机应用中,在加载游戏、更新应用等情况时,屏幕会出现一个进度栏,这里称为进度条。Android 使用 ProgressBar(进度条)控件可以完成进度条的功能。

ProgressBar 控件继承于 View 类,其常用属性如下。

android:max:进度条的最大值。

android:progress:进度条已完成进度值。

android:progressDrawable:设置轨道对应的 Drawable 对象。

android:indeterminate:如果设置成 true,则进度条不精确显示进度。

android:indeterminateDrawable:设置不显示进度条的 Drawable 对象。

android:indeterminateDuration:设置不精确显示进度的持续时间。

android:secondaryProgress:二级进度条,类似于视频播放,其中一条是当前播放进度,通过 progress 属性进行设置;另一条是缓冲进度。

style:ProgressBar 控件的样式,取值为@android:style/Widget.ProgressBar.Large(middle/small)表示大小,@android: style/Widget.ProgressBar.Horizontal 表示水平进度条,图 3.11 展示了不同样式的 ProgressBar 控件。

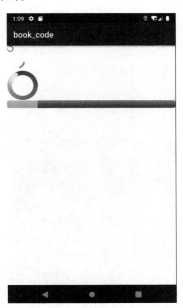

图 3.11 ProgressBar 控件

3.4 列表视图

Android 使用列表视图(ListView 控件)在同一个界面中显示多个条目,采用 MVC 设计模式的适配器类(Adapter)为其设置显示内容。

3.4.1 ListView 控件

ListView 控件以垂直方式列出列表项。ListView 控件有两个职能：一个是将数据填充到布局；另一个是处理用户的选择操作。ListView 控件的常用属性如表 3.2 所示。

表 3.2 ListView 控件的常用属性

属 性	说 明
footerDividersEnabled	是否在 footerView（表尾）前绘制一个分隔条，默认为 true
headerDividersEnabled	是否在 headerView（表头）前绘制一个分隔条，默认为 true
divider	设置分隔条，可以用颜色分割，也可以用 drawable 资源分割
dividerHeight	设置分隔条的高度
stackFromBottom	列表是否从底部开始显示
scrollbars	是否显示控件
cacheColorHint	设置单击颜色
entries	列表中的条目

在 XML 布局文件中声明 ListView 控件的语句如下：

```xml
<ListView
    android:id="@+id/list_view"
    android:layout_width="match_parent"
    android:layout_height="match_parent"/>
```

ListView 控件的使用包括添加 ListView 组件、存储数据、设置列表项 item 的布局文件、加载数据/资源进行显示、添加监听。

下面通过一个简单示例演示 ListView 控件的使用。

【例 3.6】ListView 控件的示例。

（1）创建应用程序，编辑 XML 布局文件。

创建一个应用程序，在 res/values 目录下创建资源文件 array.xml，文件内容如下：

```xml
<?xml version="1.0" encoding="utf-8"?>
<resources>
    <string-array name="languages">
        <item>c 语言</item>
        <item>java </item>
        <item>php</item>
        <item>xml</item>
        <item>html</item>
    </string-array>
</resources>
```

在主界面的布局中添加一个 ListView 控件，使用 entries 属性为其设置显示内容，代码如下：

```xml
<?xml version="1.0" encoding="utf-8"?>
<android.support.constraint.ConstraintLayout
    xmlns:android="http://schemas.android.com/apk/res/android"
    xmlns:app="http://schemas.android.com/apk/res-auto"
    xmlns:tools="http://schemas.android.com/tools"
    android:layout_width="match_parent"
    android:layout_height="match_parent"
    tools:context=".MainActivity">
    <ListView
```

```xml
        android:id="@+id/listView"
        android:layout_width="match_parent"
        android:layout_height="match_parent"
        android:entries="@array/languages" />
</android.support.constraint.ConstraintLayout>
```

（2）编写 MainActivity 代码。

为 ListView 控件添加事件监听器，该事件监听器是一个实现了 AdapterView.OnItemClickListener 接口的对象，使用时需要重写事件处理方法 onItemClick()：

```
public void onItemClick(AdapterView<?> adapterView, View view, int i, long l)
```

第一个参数代表 ListView 控件的适配器对象，第二个参数是单击的列表项对象，第三个参数是此列表项索引，第四个参数是此列表项 ID。

代码如下：

```java
public class MainActivity extends AppCompatActivity {
    @Override
    protected void onCreate(Bundle savedInstanceState) {
        super.onCreate(savedInstanceState);
        setContentView(R.layout.activity_main);
        ListView listView = (ListView) findViewById(R.id.listView);
        listView.setOnItemClickListener(new AdapterView.OnItemClickListener() {
            @Override
            public void onItemClick(AdapterView<?> adapterView, View view, int i, long l) {
                String lan[] = getResources().getStringArray(R.array.languages);
                Toast.makeText(MainActivity.this,
                        "您选择的是：" + lan[i], Toast.LENGTH_LONG).show();
            }
        });
    }
}
```

运行程序，单击列表项，在弹出的 Toast 控件中显示需要选择的内容，程序运行结果如图 3.12 所示。

图 3.12　ListView 控件

3.4.2 适配器

Android 系统使用适配器（Adapter）为 ListVIew 控件设置显示内容，是 ListView 界面与数据之间的桥梁。适配器采用如图 3.13 所示的 MVC 模型，ListView 控件相当于 V（View，视图），用于显示视图；Adapter 相当于 C（Controller，控制器），当需要数据时，ListView 会从 Adapter 中取出数据进行显示。Adapter 的工作原理如图 3.14 所示。

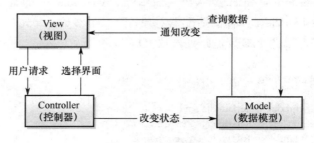

图 3.13 MVC 模型

Android 提供的主要适配器如下。

BaseAdapter：抽象类，实际开发中常继承这个类且重写相关方法，是用得最多的一个 Adapter。

ArrayAdapter：支持泛型操作，只能展现一行文字。

SimpleAdapter：同样具有良好扩展性的一个 Adapter，可以自定义多种效果。

SimpleCursorAdapter：用于显示简单文本类型的 ListView，一般在数据库操作时使用。

【例 3.7】Adapter 示例。

使用 Adapter 创建图文混排列表项，如图 3.15 所示。

图 3.14 Adapter 工作原理

图 3.15 Adapter 示例

（1）创建应用程序，编辑主界面布局文件。

创建一个应用程序，在主界面布局中添加一个 ListView 控件，代码如下：

```xml
<?xml version="1.0" encoding="utf-8"?>
<android.support.constraint.ConstraintLayout
    xmlns:android="http://schemas.android.com/apk/res/android"
    xmlns:app="http://schemas.android.com/apk/res-auto"
    xmlns:tools="http://schemas.android.com/tools"
    android:layout_width="match_parent"
    android:layout_height="match_parent"
    tools:context=".MainActivity">
    <ListView
        android:id="@+id/softList"
        android:layout_width="match_parent"
        android:layout_height="match_parent" />
</android.support.constraint.ConstraintLayout>
```

（2）创建列表项布局文件。

创建列表项布局文件 adapter.xml，使用水平 LinearLayout 布局，在其中添加 ImageView 控件和 TextView 控件，代码如下：

```xml
<?xml version="1.0" encoding="utf-8"?>
<LinearLayout xmlns:android="http://schemas.android.com/apk/res/android"
    android:layout_width="match_parent"
    android:layout_height="match_parent">
    <ImageView
        android:id="@+id/img"
        android:layout_width="100dp"
        android:layout_height="100dp" />
    <TextView
        android:id="@+id/title"
        android:layout_width="wrap_content"
        android:layout_height="wrap_content"
        android:textSize="40sp"
        android:textStyle="bold" />
</LinearLayout>
```

（3）编写 Adapter 的相关代码。

首先，创建一个名称为 Soft 的 Java Bean 类，添加属性：

```java
private int imgId;
private String title;
```

用于表示 ImageView 控件的资源 ID 和 TextView 控件中显示的内容。

然后，创建一个继承于 ArrayAdapter 的 Adapter 类 SoftAdapter，重写 getView 函数来生成列表项视图，代码如下：

```java
@Override
public View getView(int position, View convertView, ViewGroup parent) {
    Soft soft = getItem(position);
    View view = LayoutInflater.from(getContext()).inflate(resourceId, parent, false);
```

```java
ImageView imageView = (ImageView) view.findViewById(R.id.img);
TextView textView = (TextView) view.findViewById(R.id.title);
imageView.setImageResource(soft.getImgId());
textView.setText(soft.getTitle());
return view;
}
```

（4）编写 MainActivity 文件。

首先，添加 SoftAdapter 对象生成代码：

```java
int[] ids = {R.drawable.qq, R.drawable.tb, R.drawable.wx};
String titles[] = {"QQ 程序", "淘宝程序", "微信程序"};
List<Soft> softs = new ArrayList<Soft>();
for (int i = 0; i < ids.length; ++i) {
    Soft soft = new Soft(ids[i], titles[i]);
    softs.add(soft);
}
SoftAdapter adapter = new SoftAdapter(AdapterActivity.this, R.layout.adapter, softs);
```

然后，初始化 ListView 对象，挂载适配器和添加事件监听器，代码如下：

```java
ListView listView = (ListView) findViewById(R.id.softList);
listView.setAdapter(adapter);
listView.setOnItemClickListener(new AdapterView.OnItemClickListener() {
    @Override
    public void onItemClick(AdapterView<?> adapterView, View view, int i, long l) {
        Toast.makeText(AdapterActivity.this,
            "您选择的是：" + titles[i], Toast.LENGTH_SHORT).show();
    }
});
```

运行程序，可以看见图文混排自定义的列表项（见图 3.15），单击列表项可选中动作。

3.5 自定义控件

在 Android 中所有的 UI 控件都是 View 类的直接或间接子类，因此，开发人员可以继承 View 类创建自定义控件，把需要的控件组合起来变成一个新控件在以后的工程中复用。

自定义控件最重要的工作之一就是正确的绘制控件，View 的绘制由 measure()、layout()、draw()三个方法完成。

measure()方法：测量 View 的宽高。

layout()方法：设置视图在屏幕中显示的位置。

draw()方法：完成视图的绘制工作。

需要注意的是，Android 设置的坐标系是原点在视图左上角，水平为 X 轴，从左到右增大，垂直为 Y 轴，从上到下增大。

【例 3.8】自定义控件的示例。

通过自定义控件在屏幕上显示一个绿色的小球，并能随着手指在屏幕上移动。

（1）自定义 View 类。

在应用程序的包中定义一个 View 类的子类 MyView，并重写 onDraw()方法和 onTouchEvent()方法。onDraw()方法用于绘制控件，onTouchEvent()方法将每次手势移动产生的坐标设置给 MyView，并通过调用 invalidate()方法重绘控件，实现 MyView 滑动的实时更新，代码如下：

```java
public class MyView extends View {
    public    float x=100;
    public    float y=100;
    public MyView(Context context) {
        super(context);
    }
    public MyView(Context context, @Nullable AttributeSet attrs) {
        super(context, attrs);
    }
    @Override
    protected void onDraw(Canvas canvas) {
        super.onDraw(canvas);
        canvas.drawColor(Color.WHITE);
        Paint paint=new Paint();
        paint.setColor(Color.GREEN);
        canvas.drawCircle(x,y,50,paint);
    }
    @Override
    public boolean onTouchEvent(MotionEvent event) {
        x=event.getX();
        y=event.getY();
        invalidate();
        return true;
    }
}
```

（2）将自定义控件添加到 XML 布局文件中。

在 MainActivity 的 XML 布局文件中直接设置自定义控件标签，代码如下：

```xml
<?xml version="1.0" encoding="utf-8"?>
<FrameLayout xmlns:android="http://schemas.android.com/apk/res/android"
    xmlns:tools="http://schemas.android.com/tools"
    android:layout_width="match_parent"
    android:layout_height="match_parent"
    tools:context=".MainActivity">
    <!-- com.example.movedball 为自定义控件类所在的包名-->
    <com.example.movedball.MyView    com.example.movedball
      android:layout_width="wrap_content"
      android:layout_height="wrap_content"/>
</FrameLayout>
```

运行程序，显示结果如图 3.16 所示，当手指在屏幕上移动时，小球会随之移动。

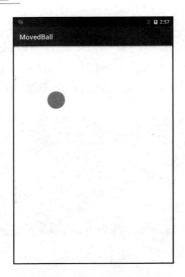

图 3.16 移动的小球

习题 3

1. 编写一个简单的选课系统。

（1）主界面上包括"填写学号和姓名""选择专业""选择课程""提交""退出"按钮，如图 3.17 所示。

（2）单击"填写学号和姓名"按钮，弹出如图 3.18 所示的自定义对话框，输入学号和姓名，单击对话框上的"确定"按钮，将输入的学号和姓名显示在主界面相应的输入编辑框内。

（3）单击"选择专业"按钮，弹出如图 3.19 所示的单选对话框，选择专业后，单击对话框上的"确定"按钮，将选择的专业显示在主界面相应的输入编辑框内。

图 3.17 主界面　　　图 3.18 自定义对话框　　　图 3.19 单选对话框

（4）单击"选择课程"按钮，弹出如图3.20所示的多选对话框，选择课程后，单击对话框上的"确定"按钮，将选择的专业显示在主界面相应的输入编辑框内。

（5）单击"提交"按钮，在下部的输入编辑框中显示选课信息，如图3.21所示。

（6）单击"退出"按钮，弹出如图3.22所示的普通对话框，单击对话框上的"确定"按钮，退出应用。

图3.20　多选对话框　　　　图3.21　显示选课信息　　　　图3.22　普通对话框

2．使用ListView控件设计一个人员信息显示界面，如图3.23所示。

图3.23　人员信息显示界面

第 4 章 程序基本单元 Activity

Activity 是 Android 的四大组件之一，用于提供可视化用户界面并与用户交互。通常一个应用程序由多个 Activity 组成，这些 Activity 之间需要经常切换并进行数据交换。一个 Activity 可以有多个 Fragment，一个 Fragment 可以被多个 Activity 重用。本章介绍 Activity 的运行状态和生命周期，Activity 的使用，多个 Activity 之间的跳转、数据传递和回传，以及 Fragment 的使用。

4.1 Activity 概述

一个 Activity 对象被赋予一个窗口，通过 setContentView(View view)方法设定要显示的视图，例如：

    setContentView(R.layout.activity_main)

一个 Activity 展示一个可视化用户界面，如果需要多个可视化用户界面，该 Android 应用会包含多个 Activity，所有的 Activity 都必须在工程的 Manifest.xml 文件中注册。在 Manifest.xml 中还需指定一个主 Activity(MainActivity)作为应用程序的启动界面，设置代码如下：

    <actionandroid:name="android.intent.action.MAIN" />

即应用程序启动时出现的第一个用户界面。

Activity 可以通过启动其他的 Activity 进行相关操作，也可以请求其他 Activity 返回的数据。当启动其他的 Activity 时，当前的 Activity 会停止，新的 Activity 会被压入栈中，同时获得用户焦点。多个 Activity 之间通过 Intent 组件进行通信。

从 MVC 设计模式的角度来说，Android 应用程序的一页其实是由 Layout XML 文件（View）与 Activity(Controller)组成的，其中的 Activity 扮演着控制页面流程的角色，是一个页面最核心的部分。

4.2 Activity 的生命周期

4.2.1 生命周期状态

Activity 的生命周期指 Activity 从创建到销毁的过程。每个 Activity 在其生命周期中最多有四种状态。

1. 运行状态（Active / Running）

Activity 处于屏幕的最前端（Activity 栈顶），它是可见、有焦点的，可以与用户进行交互。

2．暂停状态（Paused）

Activity 失去焦点，但是仍可见。例如，当前 Activity 覆盖了一个透明或非全屏的界面时，被覆盖的 Activity 就处于暂停状态。除在系统内存极端时，一个暂停状态的 Activity 依然保持活力。

3．停止状态（Stopped）

Activity 失去焦点，不可见。例如，一个 Activity 被另外的 Activity 完全覆盖，它依然保持所有状态和成员信息，但是它的窗口会因被隐藏而不再可见，系统可以随时将其释放。

4．销毁状态（Destroyed）

如果一个 Activity 是 Paused 或 Stopped 状态，系统可以将该 Activity 从内存中删除。Android 系统有两种删除方式：一种是要求该 Activity 结束；另一种是直接删除其进程。当该 Activity 再次显示给用户时，必须重新开始和重置前面的状态。

4.2.2 生命周期方法

Android 为了方便开发者能够方便地制定页面每个阶段要执行什么程序，在 Activity 类中预定义了与生命周期各阶段对应的回调方法。因此，开发人员可以选择性地重写这些方法。

1．onCreate(Bundle)

当 Activity 第一次创建时会被调用，通常做一些初始化的设置，如创建一个视图（view）、绑定列表的数据等。如果能捕获到 Activity 状态的话，这个方法传递进来的 Bundle 对象将存放 Activity 当前的状态。该方法只会被调用一次。

2．onStart()

当启动 Activity 时被调用，Activity 即将可见。

3．onResume()

当 Activity 获取焦点时被调用，将开始与用户进行交互。

4．onRestart()

在 Activity 停止后再重新启动时被调用。一般情况下，当前 Activity 从不可见重新变为可见状态，onRestart 就会被调用。

5．onPause()

当前 Activity 被其他 Activity 覆盖或屏幕锁屏时调用。如果 Activity 返回到前台将会调用 onResume()。

6．onStop()

在 Activity 对用户不可见时被调用。

7．onDestroy()

在实际开发中使用 Activity 时并不是每个方法都需要重写，而是根据需要选择重写指定的方法即可。大部分的 Activity 子类都需要实现 onCreate(Bundle)和 onPause()这两种方法。onCreate(Bundle)方法是初始化 Activity 的地方,在这里通常可以调用 setContentView(int)来设置在资源文件中定义的 UI,使用 findViewById(int)可以获得 UI 中定义的控件。onPause()方法是使用者准备离开 Activity 的地方,在这里,任何修改都应被提交。

图 4.1 显示了 Activity 的重要状态转换,矩形框表明 Activity 在状态转换之间的回调方法,椭圆框表明 Activity 所处的状态。

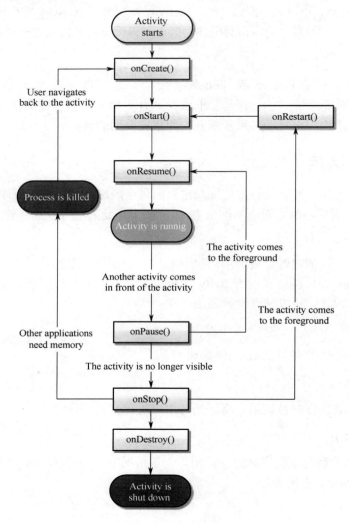

图 4.1 Activity 的生命周期

Activity 的生命周期可分为完全生命周期、可视生命周期和活动生命周期,每种生命周期中都包含不同的回调方法。

1. 完全生命周期

完全生命周期是从 Activity 创建到销毁的全部过程,从调用 onCreate(Bundle)开始到 onDestroy()结束。Activity 在 onCreate()中设置所有的全局资源和状态,并在 onDestroy()中释放这些资源。例如，某个 Activity 有一个在后台运行的线程，用于从网络下载数据，则该 Activity 可以在 onCreate()中创建线程，在 onDestroy()中停止线程。在一些极端情况下，Android 系统不调用 onDestroy()，直接终止进程。

2. 可视生命周期

可视生命周期是指 Activity 在界面上从可见到隐藏的过程，从调用 onStart()开始到 onStop()结束。onStart()用来初始化或启动与更新界面相关的资源，onStop()用来暂停或停止一切与更新用户界面相关的线程、计时器和服务。onReStart()在 onStart()前被调用，用来在 Activity 从不可见变为隐藏的过程中进行一些特定的处理过程。onStart()和 onStop()会被多次调用，使 Activity 可以随时在可见和隐藏之间转换。

3. 活动生命周期

活动生命周期是指 Activity 在屏幕的最上层，并能够与用户交互的阶段，从调用 onResume()开始到 onPause()结束。Activity 可以经常在 Resumed 和 Paused 状态之间切换，所以在这些回调方法中的代码都属于轻量级。

下面通过一个例子来演示 Activity 的生命周期。

【例 4.1】Activity 的生命周期。

首先创建应用程序 LifeofActivity，在 MainActivity 中重写 Activity 生命周期的七个回调方法，通过在回调方法中添加日志输出代码，来观察该方法的具体调用情况，程序运行后日志信息会显示在 LogCat 中。该应用程序仅演示 Activity 的生命周期，使用默认界面即可。

具体代码如下：

```java
public class MainActivity extends AppCompatActivity {
    @Override
    protected void onCreate(Bundle savedInstanceState) {
        super.onCreate(savedInstanceState);
        setContentView(R.layout.activity_main);
        Log.i("MainActivity","onCreate()被调用");        //输出日志
    }
    @Override
    protected void onStart() {
        super.onStart();
        Log.i("MainActivity","onStart()被调用");         //输出日志
    }
    @Override
    protected void onResume() {
        super.onResume();
        Log.i("MainActivity","onResume()被调用");        //输出日志
    }
    @Override
    protected void onPause() {
        super.onPause();
        Log.i("MainActivity","onPause()被调用");         //输出日志
    }
    @Override
    protected void onRestart() {
        super.onRestart();
        Log.i("MainActivity","onRestart()被调用");       //输出日志
    }
    @Override
```

```
            protected void onStop() {
                super.onStop();
                Log.i("MainActivity","onStop()被调用");        //输出日志
            }
            @Override
            protected void onDestroy() {
                super.onDestroy();
                Log.i("MainActivity","onDestroy()被调用");     //输出日志
            }
        }
```

按以下步骤运行程序，并观察 LogCat 的显示结果。

（1）运行程序，LogCat 的显示结果如图 4.2 所示。

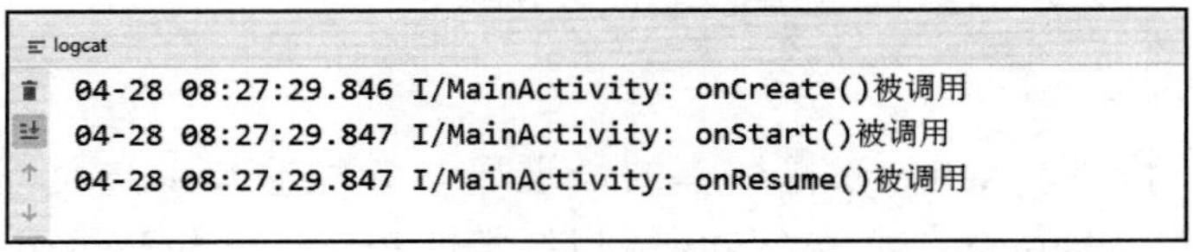

```
04-28 08:27:29.846 I/MainActivity: onCreate()被调用
04-28 08:27:29.847 I/MainActivity: onStart()被调用
04-28 08:27:29.847 I/MainActivity: onResume()被调用
```

图 4.2　启动 Activity 时的 LogCat 日志信息

通过 LogCat 日志信息可以发现，程序启动时依次调用了 onCreate()、onStart()、onResume() 这三种方法后，程序不再向下进行。在启动 Activity 时，系统首先调用 onCreate() 方法分配资源，再调用 onStart() 方法将 Activity 显示在屏幕上，然后调用 onResume() 方法获取焦点，此时应用程序处于运行状态，等待与用户进行交互。

（2）单击模拟器上的 Home 键，LogCat 中的日志信息如图 4.3 所示。

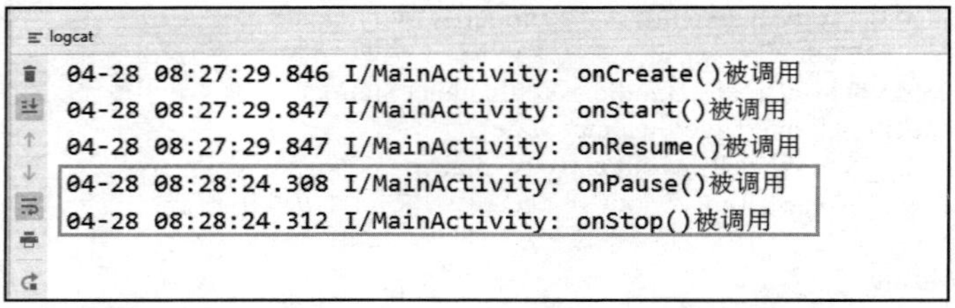

```
04-28 08:27:29.846 I/MainActivity: onCreate()被调用
04-28 08:27:29.847 I/MainActivity: onStart()被调用
04-28 08:27:29.847 I/MainActivity: onResume()被调用
04-28 08:28:24.308 I/MainActivity: onPause()被调用
04-28 08:28:24.312 I/MainActivity: onStop()被调用
```

图 4.3　按 Home 键时的 LogCat 日志信息

按 Home 键时返回到手机界面，系统依次调用 onPause() 方法和 onStop() 方法，Activity 切换至后台，此时 Activity 处于停止状态，对用户不可见。

（3）在手机界面上找到应用程序，并单击打开，LogCat 中的日志信息如图 4.4 所示。

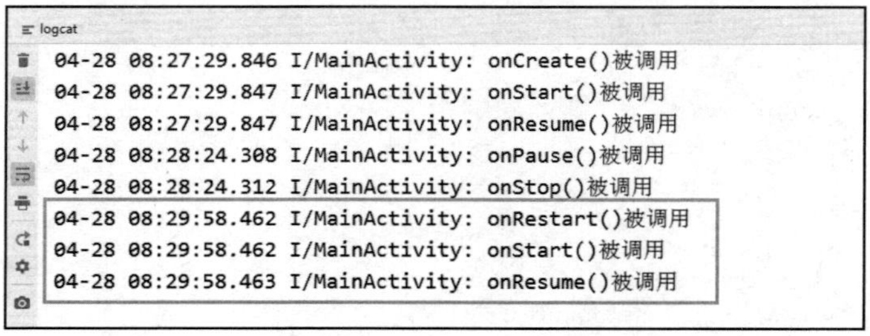

```
04-28 08:27:29.846 I/MainActivity: onCreate()被调用
04-28 08:27:29.847 I/MainActivity: onStart()被调用
04-28 08:27:29.847 I/MainActivity: onResume()被调用
04-28 08:28:24.308 I/MainActivity: onPause()被调用
04-28 08:28:24.312 I/MainActivity: onStop()被调用
04-28 08:29:58.462 I/MainActivity: onRestart()被调用
04-28 08:29:58.462 I/MainActivity: onStart()被调用
04-28 08:29:58.463 I/MainActivity: onResume()被调用
```

图 4.4　打开应用时的 LogCat 日志信息

重新打开应用程序时，依次执行 onRestart()方法、onStart()方法和 onResume()方法，Activity 重新切换到前台。

（4）单击返回键，可以看到程序已退出，同时 LogCat 中有新的日志输出，如图 4.5 所示。

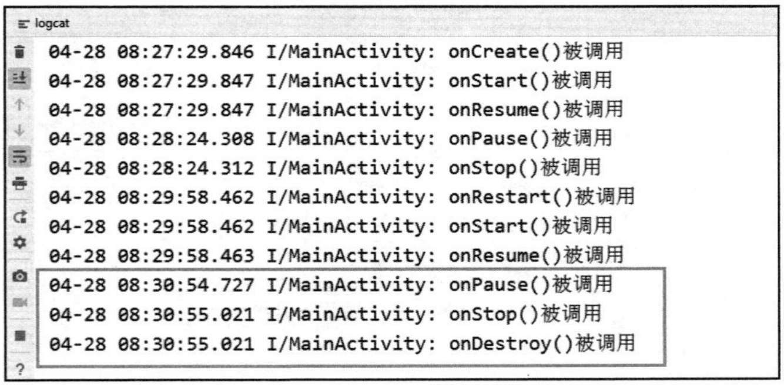

```
logcat
04-28 08:27:29.846 I/MainActivity: onCreate()被调用
04-28 08:27:29.847 I/MainActivity: onStart()被调用
04-28 08:27:29.847 I/MainActivity: onResume()被调用
04-28 08:28:24.308 I/MainActivity: onPause()被调用
04-28 08:28:24.312 I/MainActivity: onStop()被调用
04-28 08:29:58.462 I/MainActivity: onRestart()被调用
04-28 08:29:58.462 I/MainActivity: onStart()被调用
04-28 08:29:58.463 I/MainActivity: onResume()被调用
04-28 08:30:54.727 I/MainActivity: onPause()被调用
04-28 08:30:55.021 I/MainActivity: onStop()被调用
04-28 08:30:55.021 I/MainActivity: onDestroy()被调用
```

图 4.5　应用程序退出时的 LogCat 日志信息

单击返回键时系统依次调用 onPause()方法、onStop()方法和 onDestroy()方法来结束 Activity，释放资源并销毁进程。至此，整个 Activity 生命周期已完成。

分析 Activity 的生命周期，可以得出以下结论：

（1）无论 Activity 是初始启动还是暂停后重新开始，onResume()方法是 Activity 在显示前必经的阶段。因此，该阶段是屏幕数据刷新的最佳机会。

（2）无论 Activity 被隐藏还是被停止，onPause()方法是 Activity 离开显示后必经的阶段。因此该阶段是保存状态的最佳机会。

4.3　Activity 的使用

4.3.1　创建 Activity

一个应用程序所包含每个 Activity 都被定义为一个独立的类，并继承 android.app.Activity 类或其子类。Android Studio 在 API 22 之后，当创建 Android 应用时，MainActivity 会自动继承 androidx.appcompat.app.AppCompatActivity，该类是 Activity 的子类，用来替代已过时的 ActionBarActivity。AppCompatActivity 类自带标题栏，并能够兼容低版本。

在新建应用程序时，Android Studio 自动创建了一个 MainActivity。如果在应用程序中新建 Activity，通常有以下两种方式。

1．在 Android Studio 中创建 Activity

（1）在程序的包名处右击，执行"New"→"Activity"→"Empty Activity"菜单命令，如图 4.6 所示。

（2）在弹出的"Configure Activity"界面中分别填写 Activity Name（Activity 名称）、Layout Name（布局名称）和 Package name（包名），默认勾选"Generate a Layout File"复选框，用于同时创建一个布局文件，若勾选"Launcher Activity"复选框则会替代之前的启动项作为默认启动的 Activity。单击"Finish"按钮完成创建，如图 4.7 所示。

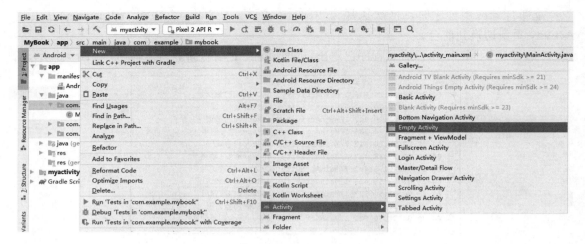

图 4.6 创建 Activity

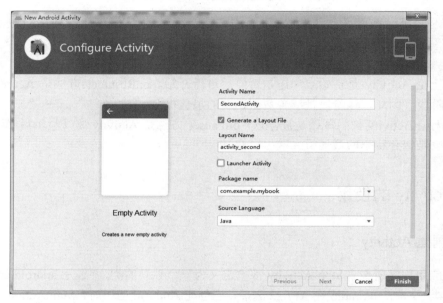

图 4.7 "Configure Activity"界面

该 Activity 对应布局文件 activity_second.xml。

创建完成的 SecondActivity 的具体代码如下：

```
package com.example.activity;
import androidx.appcompat.app.AppCompatActivity;
import android.os.Bundle;
public class SecondActivity extends AppCompatActivity {
    @Override
    protected void onCreate(Bundle savedInstanceState) {
        super.onCreate(savedInstanceState);
        setContentView(R.layout.activity_second);
    }
}
```

2. 通过继承 AppCompatActivity 类的方式创建 Activity

在应用程序包名处右击,执行"New"→"Java Class"菜单命令,弹出"New Java Class"对话框,如图 4.8 所示,输入 Activity 的名称,选择"Class"选项,然后按 Enter 键。

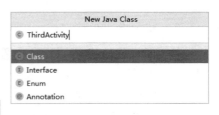

图 4.8 "New Java Class"对话框

让该类继承 AppCompatActivity 类,具体代码如下:

```
public class ThirdActivity extends AppCompatActivity {

}
```

创建一个 Activity 需要实现一种或多种方法,其中最基本的方法是 onCreate(Bundle status),它将在 Activity 被创建时回调,然后通过 setContentView(View view)方法将要显示的布局文件展示出来。

当一个 Activity 类被定义后,这个 Activity 类何时被实例化,它所包含的方法何时被调用,都由 Android 系统决定,开发者只需要实现相应的方法创建出需要的 Activity 即可。

4.3.2 配置 Activity

Android 应用要求所有应用组件(Activity、Service、ContentProvider、BroadcastReceiver)都必须先进行配置。

使用上述第一种方式创建 Activity 时,Android Studio 会自动对该 Activity 进行配置。例如,创建 SecondActivity 时,系统会在 AndroidManifest.xml 文件中自动添加如下代码用于注册 SecondActivity:

```
<activity android:name=".SecondActivity"/>
```

用上述第二种方式创建 Activity 时,需要在 AndroidManifest.xml 文件的<application>标签中,手动添加<activity>标签进行配置。

配置 Activity 时的常用属性如表 4.1 所示。

表 4.1 Activity 的常用属性

属性	说明	属性	说明
name	指定 Activity 的类名	exported	指定该 Activity 是否允许被其他应用调用
icon	指定 Activity 的对应图标	launchMode	指定 Activity 的启动模式
label	指定 Activity 的标签		

除了以上一些属性,Activity 中还可以设置一个或多个<intent-filter>标签用于指定该 Activity 可响应的 Intent。

例如,采用第二种方式创建的 ThirdActivity,就需要在 AndroidManifest.xml 文件中手动添加配置信息,否则应用程序启动时系统找不到该组件。

```
<activity  android:name="com.example.mybook.ThirdActivity"/>
```

如果 Activity 所在的包与 AndroidManifest.xml 文件的<manifest>标签中通过 package 属性指定的包名一致,则 android:name 属性值可以直接设置为".Activity 名称"。

上述创建的 ThirdActivity 的配置代码可以简化为:

```
<activity android:name=". ThirdActivity "/>
```

4.3.3 启动 Activity 和关闭 Activity

1．启动 Activity

在一个 Android 应用程序中通常会有多个 Activity，每个 Activity 都可以被其他 Activity 启动，但只有一个 Activity 作为入口，即程序启动时只会启动作为入口的 Activity，其他的会被已经启动的 Activity 启动。

启动 Activity 有以下两种方式。

（1）自动启动。

在 AndroidManifest.xml 文件中将 Activity 设置为程序的入口，当运行该程序时，系统将自动启动该 Activity。例如 MainActivity，配置信息如下：

```
<activity android:name=".MainActivity">
    <intent-filter>
        <action android:name="android.intent.action.MAIN" />
        <category android:name="android.intent.category.LAUNCHER" />
    </intent-filter>
</activity>
```

（2）使用 Intent 启动 Activity。

通过调用 Context 的 startActivity( )方法来启动 Activity。

startActivity( )方法的格式如下：

```
public void startActivity(Intent intent)
```

该方法需要一个 Intent 类型的参数。通过 Intent 对象来显式地指定所要启动的 Activity 类名，或者先在 Intent 中包含一个目标 Activity 必须执行的动作，然后调用 startActivity( )方法去查找，并启动一个与 Intent 参数匹配的 Activity。

例如，MainActivity 显式启动 SecondActivity 的代码如下：

```
Intent intent=new Intent(MainActivity.this,SecondActivity.class);
startActivity(intent);
```

关于 Intent 的用法将在 4.4 节中详细介绍。

2．关闭 Activity

通过调用 Activity 的 finish()方法来结束当前的 Activity，也可以通过调用 finishActivity()方法结束之前启动的另一个 Activity。在大多数情况下，不应使用这些方法显式结束 Activity，因为 Android 系统会管理 Activity 的生命周期，用户无须结束自己的 Activity。调用这些方法可能对预期的用户体验产生不良影响，因此只应在确实不想让用户返回此 Activity 实例时使用。

若要结束以 startActivityForResult(Intent intent,int requestCode)方法启动的 Activity，可以使用 finishActivity(requestCode)方法。

【例 4.2】Activity 的创建、配置，以及显式启动和关闭。

（1）新建应用程序。

新建一个应用程序，然后按照 4.3.1 节介绍的方法分别在应用程序中创建 SecondActivity 和 ThirdActivity。

（2）手动添加配置信息。

在 AndroidManifest.xml 文件中为 ThirdActivity 添加配置信息。MainActivity 和新创建的两个 Activity 在清单文件中的配置信息如下：

```xml
<activity android:name=".SecondActivity"></activity>     //自动配置
<activity android:name=".ThirdActivity"> </activity>     //手动配置
<activity android:name=".MainActivity">                  //自动配置
    <intent-filter>
        <action android:name="android.intent.action.MAIN" />
        <category android:name="android.intent.category.LAUNCHER" />
    </intent-filter>
</activity>
```

（3）添加控件。

在 activity_main.xml 布局文件中拖曳一个 Button 控件。activity_main.xml 代码如下：

```xml
<?xml version="1.0" encoding="utf-8"?>
<LinearLayout xmlns:android="http://schemas.android.com/apk/res/android"
    xmlns:tools="http://schemas.android.com/tools"
    android:layout_width="match_parent"
    android:layout_height="match_parent"
    android:orientation="vertical"
    tools:context=".MainActivity">
    <Button
        android:id="@+id/button_exp"
        android:layout_gravity="center_vertical"
        android:layout_width="wrap_content"
        android:layout_height="wrap_content"
        android:layout_marginTop="50dp"
        android:text="显式启动 SecondActivity"
        android:textSize="30dp" />
</LinearLayout>
```

在 activity_second.xml 布局文件中拖曳一个 TextView 控件和一个 Button 控件。activity_second.xml 的代码如下：

```xml
<?xml version="1.0" encoding="utf-8"?>
<LinearLayout xmlns:android="http://schemas.android.com/apk/res/android"
    xmlns:tools="http://schemas.android.com/tools"
    android:layout_width="match_parent"
    android:layout_height="match_parent"
    android:orientation="vertical"
    tools:context=".SecondActivity">
    <TextView
        android:id="@+id/textView"
        android:layout_width="wrap_content"
        android:layout_height="wrap_content"
        android:layout_gravity="center"
        android:layout_marginTop="30dp"
        android:textSize="30dp"
        android:text="This is SecondActivity" />
    <Button
        android:id="@+id/button"
        android:layout_width="wrap_content"
        android:layout_height="wrap_content"
```

```xml
                android:layout_gravity="center"
                android:layout_marginTop="30dp"
                android:textSize="30dp"
                android:text="Finish" />
</LinearLayout>
```

（4）编写代码。

MainActivity 的代码如下：

```java
public class MainActivity extends AppCompatActivity implements View.OnClickListener {
    private Button btn_exp;
    Intent intent;
    @Override
    protected void onCreate(Bundle savedInstanceState) {
        super.onCreate(savedInstanceState);
        setContentView(R.layout.activity_main);
        initView();
    }
    private void initView() {
        btn_exp = (Button) findViewById(R.id.button_exp);
        btn_exp.setOnClickListener(this);
    }
    @Override
    public void onClick(View v) {
        switch (v.getId()) {
            case R.id.button_exp:
                //显式启动 SecondActivity
                intent=new Intent(MainActivity.this,SecondActivity.class);
                startActivity(intent);
                break;
        }
    }
}
```

SecondActivity 的代码如下：

```java
public class SecondActivity extends AppCompatActivity {
    private Button button;
    @Override
    protected void onCreate(Bundle savedInstanceState) {
        super.onCreate(savedInstanceState);
        setContentView(R.layout.activity_second);
        button=findViewById(R.id.button);
        button.setOnClickListener(new View.OnClickListener() {
            @Override
            public void onClick(View v) {
                finish();   //结束当前的 Activity
            }
        });
    }
}
```

运行程序，单击图 4.9 中的"显式启动 SECONDACTIVITY"按钮启动 SecondActivity，页面跳转到如图 4.10 所示的 SecondActivity 对应的界面，单击"FINISH"按钮，结束当前的 SecondActivity，返回主界面。

图 4.9　主界面　　　　　　　　图 4.10　启动 SecondActivity

4.4　Intent 与 IntentFilter

Intent 是 Android 不同组件之间通信的载体，负责对应用中某次操作的动作、动作涉及的数据、附加数据进行描述，Android 根据 Intent 的描述找到对应的组件，将 Intent 传递给调用的组件，并完成组件的调用。通过 Intent 可以启动一个 Activity，也可以启动一个 Service，还可以发送一条广播消息来触发系统中的 BroadcastReceiver。

Android 三大组件之间的通信都以 Intent 为载体，封装当前组件在启动目标组件时所需要的信息，因此 Intent 通常翻译为"意图"。

4.4.1　Intent

1. Intent 的属性

Intent 就是指一个信息的捆绑包，Android 根据 Intent 所携带的信息查找要启动的组件。Intent 还携带了一些数据信息，以便于要启动的组件可以根据该数据做相应的处理。Intent 由七部分信息组成：ComponentName、Action、Category、Data、Type、Extras、Flags。

（1）ComponentName

ComponentName 是要启动的目标组件名，可以通过 Intent 的如下三种方法指定待启动组件的包名和类名。

```
setClass(Context ctx,Class<?>cls)
setClassName(Context ctx, String className)
setClassName(String pkg, String className)
```

还可以使用 Intent 的构造方法指定组件名。

```
Intent(Context packageContext, Class<?> cls)
```

例如，MainActivity 启动 SecondActivity，可使用 setClass()方法指定需要启动的组件，代码如下：

```
Intent intent=new Intent();
Intent.setClass(MainActivity.this,SecondActivity.class);
```

也可使用如下代码创建 Intent 对象：

```
Intent intent=new Intent(MainActivity.this,SecondActivity.class);
```

上述代码中，创建的 Intent 对象传入两个参数，第一个参数是 Context 对象，Activity 类是 Context 类的子类，所有的 Activity 对象都是 Context 的子对象，表示当前的 Activity。第二个参数是 Class 对象，它是被启动 Activity 类的 Class 对象。

ComponentName 是一个可选项，如果被设置，则直接使用该属性所指定的组件。

（2）Action（动作）

Action 指 Intent 要完成的动作。在 Android 中，Action 是一个字符串常量，用于描述一个 Android 应用程序的组件。在 AndroidManifest.xml 文件中使用<intent-filter>的<action>标签声明 Activity 所能接收的动作，Intent 对象使用 setAction()方法设置 Action 属性。

例如，在 AndroidManifest.xml 文件中将 ThirdActivity 的 Action 配置为字符串 "com.example. activity.START_ACTIVITY"，其中"com.example. activity"为 ThirdActivity 所在的包名。配置信息如下：

```
<activity android:name=".ThirdActivity">
    <intent-filter>
        <action android:name="com.example.activity.START_ACTIVITY"/>
        <category android:name="android.intent.category.DEFAULT"/>
    </intent-filter>
</activity>
```

如果 MainActivity 采用隐式 Intent 启动 ThirdActivity，那么在构造 Intent 对象时使用 setAction( )方法设置 Action 动作，代码如下：

```
Intent intent =new Intent();
intent.setAction("com.example. activity.START_ACTIVITY");
```

Intent 对象设置的动作要和清单文件中设置的 Action 属性值一致。

一个 Intent 对象只能指定一个 Action，但是一个 Activity（或广播等组件）可以设置（监听、匹配）多个 Action（intent-filter 中可以设置多个 Action 属性），这是两个不同的概念。

在 Intent 类中提供了大量的标准 Action 常量，其中用于启动 Activity 的标准 Action 常量及对应的字符串如表 4.2 所示。

表 4.2 常用的标准 Action 常量

Action 常量	字 符 串	描 述
ACTION_MAIN	android.intent.action.MAIN	应用程序入口
ACTION_VIEW	android.intent.action.VIEW	显示指定数据
ACTION_EDIT	android.intent.action.EDIT	请求一个 Activity，要求该 Activity 可以编辑 Intent 的 URI 中的数据
ACTION_PICK	android.intent.action.PICK	启动一个 Activity，可以从 Intent 的数据 URI 指定的 ContentProvider 中选择一个项。当关闭时，返回所选择项的 URI
ACTION_DIAL	android.intent.action.DIAL	打开一个拨号程序，要拨打的号码由 Intent 的数据 URI 预先提供。默认情况下，由本地 Android 电话拨号程序进行处理
ACTION_CALL	android.intent.action.CALL	打开一个拨号程序，并立即使用 Intent 的数据 URI 所提供的号码拨打一个电话

（续表）

Action 常量	字 符 串	描 述
ACTION_SEND	android.intent.action.SEND	启动一个 Activity，该 Activity 会发送 Intent 中指定的数据
ACTION_SENDTO	android.intent.action.SENDTO	启动一个 Activity 向 Intent 的数据 URI 所指定的联系人发送一条消息
ACTION_ANSWER	android.intent.action.ANSWER	打开一个处理来电的 Activity，通常这个动作由本地拨号程序处理

表 4.2 只列举了部分 Action 常量，读者如有兴趣可以进一步查阅 Android API 关于 Intent 的说明。

（3）Category（类别）

Category 属性用来为 Action 添加额外的附加类别信息，Category 与 Action 都是普通的字符串，通常这两个属性是结合使用的。表 4.3 中列出了 Android 系统中用于启动 Activity 的常用标准 Category 常量及对应的字符串。

表 4.3　常用的标准 Category 常量

Category 常量	字 符 串	描 述
CATEGORY_DEFAULT	android.intent.category.DEFAULT	默认的 Category
CATEGORY_BROWSABLE	android.intent.category.BROWSABLE	指定该 Activity 能被浏览器安全调用
CATEGORY_LAUNCHER	android.intent.category.LAUNCHER	Activity 显示顶级程序列表，表示在加载程序时 Activity 会出现在最上面
CATEGORY_HOME	android.intent.category.HOME	设置该 Activity 随系统启动而运行
CATEGORY_TEST	android.intent.category.TEST	该 Activity 是一个测试 Activity

（4）Data（数据）

Data 表示 Intent 对象中传递的数据。Data 属性接收一个 URI 对象，一个完整的 URI 由 scheme、host、port、path 组成，其对应的字符串格式如下：

　　　　<scheme>://<host>:<port>/<path>

scheme 是协议的名称，常见协议有 content、http、file 等，如某个图片的 URI 为 content:// media/external/images/media/4，其中 content 代表 scheme 部分，media 代表 host 部分，port 部分被省略，external/ images/media/4 代表 path 部分。

URI 就像一个数据链接，组件可以根据此 URI 获得最终的数据来源。通常将 Data 和 Action 结合使用。

例如，浏览网页：

　　　　Uri uri =Uri.parse("http://www.google.com");
　　　　Intent intent = new Intent(Intent.ACTION_VIEW,uri);
　　　　startActivity(intent);

查看联系人：

　　　　Uri personUri = Uri.parse("content://contacts/people");
　　　　Intent intent = new Intent();
　　　　intent.setAction(Intent.ACTION_PICK);
　　　　intent.setData(personUri);
　　　　startActivity(intent);

（5）Type（数据类型）

Type 属性用于 Data 属性所指定 URI 对应的 MIME 类型。它可以是任意自定义的 MIME 类型，只要是符合 abc/xyz 格式的字符串即可。

当创建一个 Intent 对象时，除了指定 URI，指定数据的 MIME 类型也很重要。例如，一个 Activity 能够显示图片，但是不能够播放视频，显示图片的 URI 和播放视频的 URI 可能很类似，为了不让 Android 误将一个含有视频 URI 的 Intent 对象传递给一个只能显示图片的 Activity，需要在该 Activity 的 Intent Filter 中指定 MIME 类型为图片，例如：

```
<data android:mimeType="image/*" />
```

并且还要给 Intent 对象设置对应的图片类型的 MIME，这样 Android 就会基于 URI 和 MIME 类型将 Intent 传递给符合条件的组件。不过有个特例，如果 URI 使用的是 Content 协议，即数据由 ContentProvider 提供，Android 就会根据 URI 自动推断出 MIME 类型，此种情况无须再指定 MIME 类型。

如果只为 Intent 对象设置数据的 URI，则需要调用 Intent 对象的 setData()方法。如果只设置数据的 MIME 类型，则需要调用 Intent 对象的 setType()方法。如果要同时设置数据的 URI 和 MIME 类型，则需要调用 Intent 对象的 setDataAndType()方法，而不是分别调用 setData()方法和 setType()方法，因为 Data 与 Type 这两个属性的设置是有顺序的，后设置的会将先设置的覆盖。例如：

- 如果为 Intent 先设置 Data 属性，再设置 Type 属性，那么 Type 属性会覆盖 Data 属性。
- 如果为 Intent 先设置 Type 属性，再设置 Data 属性，那么 Data 属性会覆盖 Type 属性。

多媒体播放：

```
Uri uri = Uri.parse("file:///sdcard/moon.mp3");
Intent intent = new Intent(Intent.ACTION_VIEW);
intent.setDataAndType(uri, "audio/mp3");
startActivity(intent);
```

（6）Extras（扩展信息）

有的 Intent 需要依靠 URI 或 Extras 携带数据的。Extras 属性是一个 Bundle 对象，在 Intent 中通过 Bundle 类型的 Extras 属性来封装数据，从而实现组件之间的数据传递。

通过调用 Intent 对象的各种重载的 putExtra(key, value)方法传入键值对形式的额外数据。

```
Intent intent =new Intent();
intent.putExtra("name","Linda") ;
intent.putExtra("age",22);
```

在被启动的组件中，通过 Intent 对象提供的 getXXXExtra()方法将传递的数据取出。

```
Intent intent=getIntent();
Striang stuName=intent.getStringExtra("name");
Int stuAge=intent.getIntExtra("age",0);
```

也可以直接创建一个 Bundle 对象，通过 Bundle 对象的各种重载的 putString(key,value)方法向该 Bundle 对象传入多个键值对，然后通过调用 Intent 对象的 putExtras(Bundle)方法将 Bundle 对象封装到 Intent 中去。

```
Bundle bundle=new Bundle();
bundle.putString("name","Linda");
bundle.putString("sex","female");
```

```
intent.putExtras(bundle);
```

在被启动的组件中，利用 getExtras()方法获取传递过来的 Bundle 对象，并通过 Bundle 对象提供的 getXXX()方法将数据取出。

```
Bundle bundle=getIntent().getExtras();
String stuName=bundle.getString("name");
int stuAge=bundle.getInt("age");
```

（7）Flags（标志位）

Flags 属性用来告知 Android 系统如何启动 Activity（如新启动的 Activity 属于哪个 task），以及在该 Activity 启动后应如何对待它。

2．Intent 的分类

根据 Intent 所描述的信息，可以将 Intent 分为两类：显式 Intent 和隐式 Intent。

（1）显式 Intent

在构造 Intent 对象时可明确指定目标组件的名称，Android 系统无须对该 Intent 做任何解析，直接找到指定的目标组件，然后启动该组件即可。

例如，在 MainActivity 中使用显式 Intent 启动 SecondActivity，可使用如下代码：

```
Intent intent=new Intent(MainActivity.this,SecondActivity.class);
startActivity(intent);
```

也可使用 setClass()方法指定需要启动的组件，代码如下：

```
Intent intent=new Intent();
Intent.setClass(MainActivity.this,SecondActivity.class);
```

在实际编程中经常使用 Intent()方法指定启动组件。

（2）隐式 Intent

在构造 Intent 对象时不用明确指定组件名，而是通过 Action、Category 和 Data 定义一个要执行的操作。Android 系统根据 Intent 指定的属性匹配 AndroidManifest.xml 中相关组件的 Intentfilter，逐一匹配出满足属性的组件。当不止一个组件满足时，会弹出一个对话框进行选择。

在构建一个隐式 Intent 时，需要指定一个所要执行的动作，除此之外，还可以提供执行该动作所需要的数据 URI。通过向 Intent 添加 Extras 的方式向目标 Activity 发送额外的数据。

隐式 Intent 启动 Activity 时，需要指定其对应的 Category（类别）、Action（动作）、Data（数据），Android 系统根据 Intent 中的数据信息，找到需要启动的 Activity。

隐式 Intent 不但可以启动自己应用的 Activity，也可启动系统的或其他应用的 Activity，如浏览器、电话拨号、通讯录、发短信、发邮件等。

① 启动浏览器打开一个网页。

```
Uri uri = Uri.parse("http://www.sohu.com");
Intent intent=new Intent(Intent.ACTION_VIEW, uri);
startActivity(intent);
```

② 启动电话拨号程序。

```
Uri uri = Uri.parse("tel:+1234567");
Intent intent =new Intent(Intent.ACTION_DIAL，uri);
startActivity(intent);
```

③ 启动联系人。

```
Uri uri = Uri.parse("content://contacts/people ");
Intent intent =new intent(Intent.ACTION_PICK,uri);
```

```
        startActivity(intent);
```
④ 发送电子邮件。
```
        Uri uri = Uri.parse("mailto:qst@163.com ");
        Intent intent =new Intent(Intent.ACTION_SENDTO，uri);
        startActivity(intent);
```

4.4.2　IntentFilter

在 Android 中，IntentFilter 表示 Intent 的过滤器，用来描述指定的组件可以处理哪些 Intent。当有 Intent 发送时，Android 根据 IntentFilter 的配置信息，找到可以处理该 Intent 的组件，并把 Intent 传递给它。Activity、Service 和 BroadcastReceiver 只有设置了 IntentFilter，才能被隐式 Intent 调用。

Activity 在 AndroidManifest.xml 文件中设置 IntentFilter。在<activity>标签下定义<intent-filter>子节点。<intent-filter>节点可支持<action>标签、<category>标签和<data>标签，分别用来定义 IntentFilter 的"动作""类别""数据"。

1．<action>标签

声明该组件可以执行的操作，action 值是关于操作的一个字符串。<intent-filter>标签中可以设置多个 action 属性，当使用隐式 Intent 激活组件时，只要 Intent 携带的 action 与<intent-filter>标签中声明的任何一个 action 相同，action 属性就匹配成功。

在 AndroidManifest.xml 文件中声明 action 属性的代码如下：
```
    <intent-filter>
        …
        <action android:name="com.example. activity.START_ACTIVITY"/>
        <action android:name="android.intent.action.EDIT"/>
        …
    </intent-filter>
```

2．<category >标签

声明接收 Intent 的类型。一个 IntentFilter 可以不声明 category 属性，也可以声明多个 category 属性。

在 AndroidManife st.xml 文件中声明 category 属性的代码如下：
```
    <intent-filter>
        …
        <category android:name="android.intent.category.DEFAULT"/>
        <category android:name="android.intent.category.BROWSABLE"/>
        …
    </intent-filter>
```

隐式 Intent 中声明的 category 属性必须全部与某一个 IntentFilter 中的 category 属性相同才算匹配成功。需要注意的是，IntentFilter 中罗列的 category 属性数量必须大于或等于隐式 Intent 携带的 category 属性数量时，才算匹配成功。如果一个隐式 Intent 没有设置 category 属性，则可以通过任何一个 IntentFilter 的 category 属性匹配。

需要注意的是，使用隐式 Intent 启动 Activity 时，系统会默认为该 Intent 对象添加"android.intent.category.DEFAULT"的 category 属性，因此为了使 Activity 能够接收隐式 Intent，必须在清单文件中为该 Activity 指定 category 属性为"android.intent.category.DEFAULT"。

3. <data>标签

<data>标签可以指定数据的 URI 和 MIME 类型，使用一个或多个<data>标签来指定组件可以接受的数据类型。一个 IntentFilter 可以罗列多个 data 属性，每个属性可以指定数据的 URI 或 MIME 类型。隐式 Intent 携带的 data 数据只要与 IntentFilter 中的任意一个 data 相同，data 属性就匹配成功。

IntentFilter 检测隐式 Intent 的 action、data、category 等属性，其中任何一项失败，Android 系统都不会传递 Intent 给此组件。一个组件可以有多个 IntentFilter，Intent 只要通过其中的某个 IntentFilter 检测，就可以调用此组件。

在隐式 Intent 启动方式中，调用者通过 action、data、category 等属性描述 Intent，被调用者通过在 AndroidManifest.xml 文件中声明的一系列 IntentFilter 来描述自己能够响应哪些 Intent。隐式 Intent 启动的优点是不需要指明启动哪个组件，而由系统来决定，这样有利于降低组件之间的耦合度，提高 Android 组件的可复用性。

通过显式 Intent 启动目标组件时，被启动的组件不需要配置 IntentFilter 就能启动。

【例 4.3】IntentFilter 的配置和隐式 Intent 启动 Activity。

本例题在【例 4.2】的基础上实现隐式 Intent 启动 ThirdActivity。

（1）为 ThirdActivity 添加<intent-filter>过滤器。

在 AndroidManifest.xml 文件中为 ThirdActivity 添加<intent-filter>过滤器，代码如下：

```xml
<activity android:name=".ThirdActivity">
    <intent-filter>
        <action android:name="com.example.activity.START_ACTIVITY"/>
        <category android:name="android.intent.category.DEFAULT"/>
    </intent-filter>
</activity>
```

（2）添加控件。

向 activity_main.xml 文件中添加一个 Button 控件，代码如下：

```xml
<Button
    android:id="@+id/button_imp"
    android:layout_gravity="center_vertical"
    android:layout_width="wrap_content"
    android:layout_height="wrap_content"
    android:layout_marginTop="50dp"
    android:text="隐式启动 ThirdActivity"
    android:textSize="30dp" />
```

（3）创建 activity_third.xml 布局文件。

在 res 文件夹下创建 activity_third.xml 布局文件，向其中拖曳一个 Button 控件和一个 TextView 控件。activity_third.xml 代码如下：

```xml
<?xml version="1.0" encoding="utf-8"?>
<LinearLayout xmlns:android="http://schemas.android.com/apk/res/android"
    xmlns:tools="http://schemas.android.com/tools"
    android:layout_width="match_parent"
    android:layout_height="match_parent"
    android:orientation="vertical"
    tools:context=". ThirdActivity ">
    <TextView
```

```xml
        android:id="@+id/textView"
        android:layout_width="wrap_content"
        android:layout_height="wrap_content"
        android:layout_gravity="center"
        android:layout_marginTop="30dp"
        android:textSize="30dp"
        android:text="This is ThirdActivity" />
    <Button
        android:id="@+id/button"
        android:layout_width="wrap_content"
        android:layout_height="wrap_content"
        android:layout_gravity="center"
        android:layout_marginTop="30dp"
        android:textSize="30dp"
        android:text="Finish" />
</LinearLayout>
```

（4）编写 ThirdActivity 代码。

```java
public class ThirdActivity    extends AppCompatActivity {
    private Button button ;
    @Override
    protected void onCreate(Bundle savedInstanceState) {
        super.onCreate(savedInstanceState);
        setContentView(R.layout.activity_third);
        button=findViewById(R.id.button);
        button.setOnClickListener(new View.OnClickListener() {
            @Override
            public void onClick(View v) {
                finish();   //结束当前的 Activity
            }
        });
    }
}
```

（5）编写 MainActivity 代码。

```java
public class MainActivity extends AppCompatActivity implements View.OnClickListener {
    private Button btn_exp;
    private Button btn_imp;
    Intent intent;
    @Override
    protected void onCreate(Bundle savedInstanceState) {
        super.onCreate(savedInstanceState);
        setContentView(R.layout.activity_main);
        initView();
    }
    private void initView() {
        btn_exp = (Button) findViewById(R.id.button_exp);
        btn_imp = (Button) findViewById(R.id.button_imp);
        btn_exp.setOnClickListener(this);
        btn_imp.setOnClickListener(this);
    }
    @Override
```

```java
            public void onClick(View v) {
                switch (v.getId()) {
                    case R.id.button_exp:
                        //显式启动 SecondActivity
                        intent=new Intent(MainActivity.this,SecondActivity.class);
                        startActivity(intent);
                        break;
                    case    R.id.button_imp:
                        //隐式启动 ThirdActivity
                        intent=new Intent();
                        //设置 Action 动作,与清单中的<action>属性一致
                        intent.setAction("com.example.activity.START_ACTIVITY");
                        startActivity(intent);
                        break;
                }
            }
        }
```

运行程序,在图 4.11 中单击"隐式启动 THIRDACTIVITY"按钮,启动 ThirdActivity,如图 4.12 所示,单击"FINISH"按钮结束 ThirdActivity,返回主界面。

图 4.11　主界面　　　　　图 4.12　启动 ThirdActivity

4.5　多个 Activity 的使用

一个 Android 应用程序通常包含多个 Activity,这些 Activity 之间需要经常切换并进行数据交换。Activity 的跳转与数据传递主要通过 Intent 实现,在启动新的 Activity 时,将要传递的数据封装在 Intent 中,然后通过调用 startActivity(Intent intent)方法或 startActivityForResult (Intent intent, int requestCode)方法将 Intent 传递给新启动的 Activity。前者直接启动一个 Activity,不用等待返回;后者启动一个 Activity 并等待回传结果。如果要启动的 Activity 存在,则 requestCode 的值将会传递到 onActivityResult 方法中。

4.5.1 Activity 之间数据的传递

利用 Intent 向目标 Activity 传递数据有三种方式。

1. 利用 putExtra()方法传递单个数据

通过 putExtra()方法将传递的数据存储在 Intent 对象中。Android 提供了多个重载的 putExtra()方法用于传递不同类型的数据。putExtra(String,value)方法的第一个参数为 String 类型的键名，第二个参数为键对应的值，可以是任何类型。

```
Intent intent =new Intent();
intent.putExtra("name","Linda") ;
intent.putExtra("age",22);
```

在目标 Activity 中，通过 getXXXExtra()方法获取 Intent 中的数据。

```
Intent intent=getIntent();    //获取 Intent 对象
String stuName=intent.getStringExtra("name");    //获取 Intent 对象中封装的数据
int stuAge=intent.getIntExtra("age",0);
```

2. 使用 Bundle 批量传递数据

Bundle 以 key-value（键值对）的形式传递数据，在 Android 开发中十分常见。Bundle 传递的数据包括 String、Int、Boolean、Byte、Float、Long、Double 等基本类型或其对应数组，也可以是对象或对象数组。当 Bundle 传递对象或对象数组时，必须实现 Serializable 接口或 Parcelable 接口。

首先通过 Bundle 对象的 putString(key, value)方法向该 Bundle 中传入多个键值对，然后通过 Intent 提供的 putExtras()方法将 Bundle 对象封装到 Intent 中。

```
Bundle bundle=new Bundle();
bundle.putString("name","Linda");
bundle.putString("age",22);
intent.putExtras(bundle);
//intent.putExtras("message",bundle);
```

在目标 Activity 中，利用 getExtras()方法获取传递过来的 Bundle 对象，并通过 Bundle 提供的 getXXX()方法将数据取出。

```
Bundle bundle=getIntent().getExtras();
//Bundle bundle=getIntent().getBundleExtras("message");
String stuName=bundle.getString("name");
int stuAge=bundle.getInt("age");
```

3. 使用 Intent 传递 Object 对象

Android 开发中有时需要在应用中或进程间传递对象，通过 Intent 传递 Object 对象有以下两种方法。

（1）Bundle.putSerializable(Key,Object)方法。需要类实现 Serializable 接口，将一个对象转换成可存储或可传输的状态。Serializable 接口序列化后的对象可以在网络上进行传输，也可以存储到本地。

（2）Bundle.putParcelable(Key,Object)方法。需要类实现 Parcelable 接口，将一个完整的对象进行分解，而分解后的每一部分都是 Intent 所支持的数据类型。

Serializable 接口是 JavaSE 本身就支持的，而 Parcelable 接口是 Android 特有的功能，要比实现 Serializable 接口高效，可用于 Intent 的数据传递，也可以用于进程间通信（IPC）。实现 Serializable 接口非常简单，声明一下就可以了，而实现 Parcelable 接口则稍微复杂一些，但效率更高，推荐用这种方法来提高性能。

通过 Intent 的 putExtra(String name,Serializable value)方法，将对象传递过去。在接收方的 Activity 中通过 intent.getExtras(name)方法获得该对象。

【例 4.4】Activity 之间数据传递。

本例通过一个登录程序演示 Activity 之间数据的传递。用户成功登录后系统将用户名传递到下一个界面并显示。

（1）登录界面设计。

新建一个应用程序，Activity 命名为 LoginActivity，对应的布局文件为 activity_login.xml，登录界面设计如图 4.13 所示。

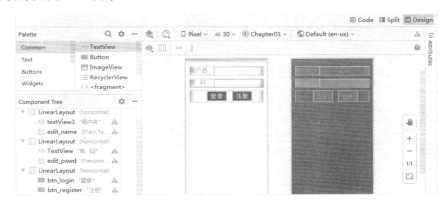

图 4.13　登录界面设计

activity_login.xml 文件的代码如下：

```xml
<?xml version="1.0" encoding="utf-8"?>
<LinearLayout xmlns:android="http://schemas.android.com/apk/res/android"
    xmlns:tools="http://schemas.android.com/tools"
    android:layout_width="match_parent"
    android:layout_height="match_parent"
    android:padding="30dp"
    android:orientation="vertical"
    tools:context=".LoginActivity">
    <LinearLayout
        android:layout_width="match_parent"
        android:layout_height="wrap_content"
        android:layout_marginTop="20dp"
        android:orientation="horizontal">
        <TextView
            android:layout_width="0dp"
            android:layout_height="wrap_content"
            android:layout_weight="1"
            android:textSize="30dp"
            android:text="用户名" />
```

```xml
        <EditText
            android:id="@+id/edit_name"
            android:layout_width="0dp"
            android:layout_height="wrap_content"
            android:layout_weight="2"
            android:ems="10"
            android:hint="请输入用户名"
            android:inputType="textPersonName"
            android:text="" />
    </LinearLayout>
    <LinearLayout
        android:layout_width="match_parent"
        android:layout_height="wrap_content"
        android:layout_marginTop="20dp"
        android:orientation="horizontal">
        <TextView
            android:layout_width="0dp"
            android:layout_height="wrap_content"
            android:layout_weight="1"
            android:textSize="30dp"
            android:text="密码" />
        <EditText
            android:id="@+id/edit_pswd"
            android:layout_width="0dp"
            android:layout_height="wrap_content"
            android:layout_weight="2"
            android:hint="请输入密码"
            android:ems="10"
            android:inputType="numberPassword" />
    </LinearLayout>
    <LinearLayout
        android:layout_width="match_parent"
        android:layout_height="match_parent"
        android:layout_marginTop="20dp"
        android:orientation="horizontal">
        <Button
            android:id="@+id/btn_login"
            android:layout_width="wrap_content"
            android:layout_height="wrap_content"
            android:layout_weight="1"
            android:textSize="30dp"
            android:text="登录" />
        <Button
            android:id="@+id/btn_register"
            android:layout_width="wrap_content"
            android:layout_height="wrap_content"
            android:layout_weight="1"
            android:textSize="30dp"
            android:text="注册" />
    </LinearLayout>
</LinearLayout>
```

（2）用户界面设计。

在应用程序中创建一个 Activity，命名为 UserActivity，对应的布局文件为 activity_user.xml，代码如下：

```xml
<?xml version="1.0" encoding="utf-8"?>
<LinearLayout xmlns:android="http://schemas.android.com/apk/res/android"
    xmlns:tools="http://schemas.android.com/tools"
    android:layout_width="match_parent"
    android:layout_height="match_parent"
    android:padding="20dp"
    tools:context=".UserActivity">
    <TextView
        android:id="@+id/tv_hello"
        android:layout_width="349dp"
        android:layout_height="65dp"
        android:layout_marginTop="30dp"
        android:text="TextView"
        android:textSize="25dp" />
</LinearLayout>
```

（3）编写"登录"按钮的事件处理代码。

为登录界面上的"登录"按钮编写事件处理代码，在页面跳转时将输入的用户名通过 Intent 传递到用户界面。

在 LoginActivity.java 中实现登录验证、页面跳转和数据传递，代码如下：

```java
public class LoginActivity extends AppCompatActivity implements View.OnClickListener {
    private EditText edName,edPswd;
    private Button btnLogin,btnRegister;
    @Override
    protected void onCreate(Bundle savedInstanceState) {
        super.onCreate(savedInstanceState);
        setContentView(R.layout.activity_login);
        initView();
    }
    public void initView(){
        edName=findViewById(R.id.edit_name);
        edPswd=findViewById(R.id.edit_pswd);
        btnLogin=findViewById(R.id.btn_login);
        btnRegister=findViewById(R.id.btn_register);
        btnLogin.setOnClickListener(this);
        btnRegister.setOnClickListener(this);
    }
    @Override
    public void onClick(View v) {
        switch (v.getId()){
            case R.id.btn_login:
                String userName=edName.getText().toString();
                String password=edPswd.getText().toString();
                if (userName.isEmpty()) {
                    Toast.makeText(LoginActivity.this, "请输入用户名",
                            Toast.LENGTH_LONG). show();
                    return;
```

```java
                    }
                    if (password.isEmpty()){
                        Toast.makeText(LoginActivity.this, "请输入密码",
                                Toast.LENGTH_LONG). show();
                        return;
                    }
                    if(userName.equals("admin") && password.equals("123")) {
                        Intent intent = new Intent(LoginActivity.this, UserActivity.class);
                        //将 userName 封装到 intent
                        intent.putExtra("name", userName);
                        startActivity(intent); //启动 UserActivity
                    }
                    else {
                        Toast.makeText(LoginActivity.this,"用户名或密码错误",
                                Toast. LENGTH_LONG ).show();
                    }
                    break;
                case   R.id.btn_register:
                    //【例 4.5】实现该按钮的事件处理
                    break;
            }
        }
    }
```

（4）处理传递过来的数据。

在 UserActivity 的 onResume()方法中接收并解析从 LoginActivity 传递过来的 Intent 对象，得到用户登录时填写的用户名，将其显示在用户界面上。UserActivity.java 代码如下：

```java
public class UserActivity extends AppCompatActivity {
    private TextView loginInfo;
    @Override
    protected void onCreate(Bundle savedInstanceState) {
        super.onCreate(savedInstanceState);
        setContentView(R.layout.activity_user);
        loginInfo=findViewById(R.id.tv_hello);
    }
    @Override
    protected void onResume() {
        super.onResume();
        //获取传递过来的 Intent
        Intent intent=getIntent();
        //从 Intent 中获取数据
        String name=intent.getStringExtra("name");
        loginInfo.setText("您好,"+name+"，欢迎光临！");
    }
}
```

运行程序，出现登录界面时，输入用户名和密码，如图 4.14 所示。单击"登录"按钮，如果输入的用户名正确，则会跳转到用户界面，显示欢迎信息，如图 4.15 所示。

图 4.14　登录界面　　　　图 4.15　用户界面

4.5.2　Activity 之间数据的回传

在开发 Android 应用时，有时需要在一个 Activity 中调用另一个 Activity。在新启动的 Activity 结束操作后，程序自动返回到原 Activity 并显示操作结果。通过 startActivity()方法启动的 Activity 与原 Activity 相互独立，在关闭时不会返回任何信息。当需要返回数据时，可以使用 startActivityForResult()方法启动一个 Activity，那么新启动的 Activity 在关闭时，将通过 onActivityResult()事件处理方法来返回结果。

1．通过 startActivityForResult( )方法启动 Activity

startActivityForResult(Intent intent, int requestCode)

第一个参数 Intent 用于显式或隐式决定启动哪个 Activity。

第二个参数 requestCode 用于标识请求的来源。如果≥0，当 Activity 结束时 requestCode 将归还到 onActivityResult()中，以便确定数据是从哪个 Activity 中返回的。

下述代码用于显式 Intent 启动一个目标 Activity，并设置相应的请求标识码为 1。

```
Intent intent=new Intent(MainActivity.this,SecondActivity.class);
startActivityForResult (intent,1);
```

下述代码用于隐式 Intent 启动一个目标 Activity 来选取联系人，并设置相应的请求标识码为 2。

```
Uri personUri = Uri.parse("content://contacts/people");
Intent intent = new Intent(Intent.ACTION_PICK, personUri);
startActivityForResult (intent,2);
```

2．目标 Activity 调用 setResult()方法向原 Activity 返回结果

setResult(int resultCode,Intent data)

第一个参数 resultCode 用于设置目标 Activity 以何种方式返回，当 Activity 结束时 resultCode 将归还到 onActivityResult()中。resultCode 的值一般为 RESULT_CANCELED 或 RESULT_OK，在某些环境下，当 CANCELED 或 OK 不足以精确描述返回结果时，用户可以使用自己的响应码来处理应用程序的特定选择，setResult()方法的 resultCode 支持使用其他任意的指数值。

第二个参数 data 是目标 Activity 所要返回的 Intent 数据载体。Intent 作为结果返回时，通常包含某段内容（如选择的联系人、电话号码或媒体文件等）的 URI 和用于返回的一组附加信息。

目标 Activity 调用 setResult()方法后，原 Activity 并不会立刻回调 onActivityResult()方法，只有当目标 Activity 调用 finish()方法后才会调用 onActivityResult()方法。

当用户通过返回键关闭 Activity，或者在调用 finish()方法之前没有调用 setResult()方法时，resultCode 将被设置为 RESULT_CANCELED，且返回结果为空。所以，setResult()方法应在 finish()方法前调用，否则会出现还没调用 setResult()方法就返回到前一个 Activity 的情况。

3．处理从目标 Activity 返回的结果

当目标 Activity 关闭时，触发并调用 Activity 的 onActivityResult()事件处理方法。通过在原 Activity 中重写 onActivityResult()方法来处理从目标 Activity 返回的结果。该方法的语法格式如下：

```
onActivityResult(int requestCode, int resultCode, Intent data)
```

第一个参数 requestCode 是在启动目标 Activity 时使用的请求码，用来确认数据是从哪个 Activity 返回的。

第二个参数 resultCode 是从目标 Activity 返回的响应码，通过其 setResult()方法返回。

第三个参数 data 是响应码对应的返回数据。根据目标 Activity 的不同，data 可能会包含代表选定内容的 URI。另外，目标 Activity 也可以通过 Intent 的 Extra 形式返回数据。

当调用 startActivityForResult()方法启动多个 Activity 时，每个 Activity 返回的数据都会回调到 onActivityResult()方法中，因此，先要通过检查 requestCode 的值来判断数据来源，然后再通过 resultCode 的值来判断数据处理结果是否成功，最后从 Data 中取出数据。

例如，Activity01 启动 Activity02，关闭 Activity02 时将向 Activity01 返回数据。操作过程如下：

（1）在 Activity01 中调用 startActivityForResult()启动 Activity01。

```
Intent intent = new Intent(this,Activity02.class);
startActivityForResult(intent,1);   //请求码为 1
```

（2）在 Activity02 中返回数据。

```
Intent intent = new Intent();
intent.putExtra("extra_data","Hello Activity01");
setResult(2,intent);   //响应码为 2
finish( ) ;
```

（3）在 Activity01 中重写 onActivityResult()方法获取返回的数据。

```
protected void onActivityResult(int requestCode, int resultCode, Intent data)  {
        super.onActivityResult(requestCode, resultCode, data);
        if (requestCode == 1){
            if (resultCode == 2) {
                String string= data.getStringExtra("extra_data");
            }
        }
    }
```

【例 4.5】Activity 之间数据的传递和回传。

在【例 4.4】的基础上增加一个注册界面和注册信息显示界面，用户单击登录界面（见图 4.16）的"注册"按钮时，跳转到用户注册界面（见图 4.17），填写用户注册信息（见图 4.18），单击"提交"按钮，跳转到注册信息界面（见图 4.19），单击该界面上的"关闭"按钮，返回用户注册界面。单击用户注册界面上的"返回登录界面"按钮，返回到登录界面，同时将注册的用户名和密码传递到登录界面（见图 4.20）。本例的数据传递在 LoginActivity（登录界面）、RegActivity（用户注册界面）和 ShowActivity（注册信息界面）之间进行。

图 4.16　登录界面（1）

图 4.17　用户注册界面

图 4.18　填写用户注册信息

图 4.19　注册信息界面

图 4.20　登录界面（2）

（1）注册界面设计。

在应用程序中创建一个 Activity，命名为 RegActivity，对应布局文件 activity_reg.xml 的代码如下：

```
<?xml version="1.0" encoding="utf-8"?>
```

```xml
<LinearLayout xmlns:android="http://schemas.android.com/apk/res/android"
    xmlns:app="http://schemas.android.com/apk/res-auto"
    xmlns:tools="http://schemas.android.com/tools"
    android:layout_width="match_parent"
    android:layout_height="match_parent"
    android:padding="10dp"
    android:orientation="vertical"
    tools:context=".RegActivity">
    <LinearLayout
        android:layout_width="match_parent"
        android:layout_height="wrap_content"
        android:layout_marginTop="20dp"
        android:orientation="horizontal">
        <TextView
            android:layout_width="0dp"
            android:layout_height="wrap_content"
            android:layout_weight="1"
            android:textSize="30dp"
            android:text="用户名" />
        <EditText
            android:id="@+id/edit_name"
            android:layout_width="0dp"
            android:layout_height="wrap_content"
            android:layout_weight="2"
            android:ems="10"
            android:hint="请输入用户名"
            android:inputType="textPersonName"
            android:text="" />
    </LinearLayout>
    <LinearLayout
        android:layout_width="match_parent"
        android:layout_height="wrap_content"
        android:layout_marginTop="20dp"
        android:orientation="horizontal">
        <TextView
            android:layout_width="0dp"
            android:layout_height="wrap_content"
            android:layout_weight="1"
            android:textSize="30dp"
            android:text="密码" />
        <EditText
            android:id="@+id/edit_pswd"
            android:layout_width="0dp"
            android:layout_height="wrap_content"
            android:layout_weight="2"
            android:hint="请输入密码"
            android:ems="10"
            android:inputType="numberPassword" />
    </LinearLayout>
    <LinearLayout
        android:layout_width="match_parent"
```

```xml
        android:layout_height="wrap_content"
        android:layout_marginTop="20dp"
        android:orientation="horizontal">
        <TextView
            android:layout_width="0dp"
            android:layout_height="wrap_content"
            android:layout_weight="1"
            android:textSize="30dp"
            android:text="确认密码" />
        <EditText
            android:id="@+id/edit_confirmpswd"
            android:layout_width="0dp"
            android:layout_height="wrap_content"
            android:layout_weight="2"
            android:hint="请再次输入密码"
            android:ems="10"
            android:inputType="numberPassword" />
    </LinearLayout>
    <LinearLayout
        android:layout_width="match_parent"
        android:layout_height="wrap_content"
        android:layout_marginTop="20dp"
        android:orientation="horizontal">
        <TextView
            android:layout_width="0dp"
            android:layout_height="wrap_content"
            android:layout_weight="1"
            android:textSize="30dp"
            android:text="邮箱" />
        <EditText
            android:id="@+id/edit_email"
            android:layout_width="0dp"
            android:layout_height="wrap_content"
            android:layout_weight="2"
            android:hint="请输入邮箱"
            android:ems="10"
            android:inputType="textEmailAddress" />
    </LinearLayout>
    <LinearLayout
        android:layout_width="match_parent"
        android:layout_height="wrap_content"
        android:layout_marginTop="20dp"
        android:orientation="horizontal">
        <TextView
            android:layout_width="0dp"
            android:layout_height="wrap_content"
            android:layout_weight="1"
            android:textSize="30dp"
            android:text="手机号码" />
        <EditText
            android:id="@+id/edit_phone"
```

```xml
            android:layout_width="0dp"
            android:layout_height="wrap_content"
            android:layout_weight="2"
            android:hint="请输入手机号码"
            android:ems="10"
            android:inputType="textPhonetic" />
    </LinearLayout>
    <LinearLayout
        android:layout_width="match_parent"
        android:layout_height="match_parent"
        android:layout_marginTop="20dp"
        android:orientation="horizontal">
        <Button
            android:id="@+id/btn_submit"
            android:layout_width="wrap_content"
            android:layout_height="wrap_content"
            android:layout_weight="1"
            android:textSize="30dp"
            android:text="提交" />
        <Button
            android:id="@+id/btn_return"
            android:layout_width="wrap_content"
            android:layout_height="wrap_content"
            android:layout_weight="1"
            android:textSize="30dp"
            android:text="返回登录界面" />
    </LinearLayout>
</LinearLayout>
```

（2）注册信息界面设计。

在应用程序中创建一个 Activity，命名为 ShowActivity，对应布局文件 activity_show.xml 的代码如下：

```xml
<?xml version="1.0" encoding="utf-8"?>
<LinearLayout xmlns:android="http://schemas.android.com/apk/res/android"
    xmlns:tools="http://schemas.android.com/tools"
    android:layout_width="match_parent"
    android:layout_height="match_parent"
    android:orientation="vertical"
    android:padding="20dp"
    tools:context=".RegisterInfoActivity">
    <TextView
        android:id="@+id/textView4"
        android:layout_width="match_parent"
        android:layout_height="wrap_content"
        android:textSize="30dp"
        android:gravity="center"
        android:text="用户注册信息" />
    <TextView
        android:id="@+id/tv_infomation"
        android:layout_width="match_parent"
```

```xml
            android:layout_height="wrap_content"
            android:text="注册信息"
            android:textSize="25dp" />
    <Button
            android:id="@+id/button"
            android:layout_width="wrap_content"
            android:layout_height="wrap_content"
            android:layout_gravity="center"
            android:layout_marginTop="20dp"
            android:text="关闭" />
</LinearLayout>
```

(3) 编写"注册"按钮的事件处理代码。

为登录界面上的"注册"按钮编写事件处理代码，单击时跳转到注册界面。在 LoginActivity.java 中添加如下代码：

```java
case R.id.btn_register:
    Intent intent=new Intent(LoginActivity.this,RegActivity.class);
    startActivityForResult(intent, 101);
    break;
```

RegActivity 结束时需要向 LoginActivity 返回数据，因此使用 startActivityForResult()方法启动 RegActivity。

(4) 在 RegActivity 中实现数据传递和回传。

单击用户注册界面的"提交"按钮时，跳转到注册信息界面，在该界面显示用户注册信息，单击"关闭"按钮，返回用户注册界面；单击"返回登录界面"按钮，返回登录界面。如果操作成功则将用户名和密码回传给 LoginActivity。RegActivity.java 代码如下：

```java
public class RegActivity extends AppCompatActivity implements View.OnClickListener {
    private Button btnSubmit,btnReturn;
    private EditText etName,etPassword,edConfirmPassword,etEmail,etPhone;
    @Override
    protected void onCreate(Bundle savedInstanceState) {
        super.onCreate(savedInstanceState);
        setContentView(R.layout.activity_reg);
        setTitle("用户注册");
        initView();
    }
    public    void initView(){
        btnSubmit=findViewById(R.id.btn_submit);
        btnReturn=findViewById(R.id.btn_return);
        etName=findViewById(R.id.edit_name);
        etPassword=findViewById(R.id.edit_pswd);
        edConfirmPassword=findViewById(R.id.edit_confirmpswd);
        etEmail=findViewById(R.id.edit_email);
        etPhone=findViewById(R.id.edit_phone);
        btnSubmit.setOnClickListener(this);
        btnReturn.setOnClickListener(this);
    }
    @Override
    public void onClick(View v) {
        String userName=etName.getText().toString();
        String userPsd=etPassword.getText().toString();
        String confirPsd=edConfirmPassword.getText().toString();
```

```java
                        String userEmail=etEmail.getText().toString();
                        String userPhone=etPhone.getText().toString();
                        switch (v.getId()){
                            case R.id.btn_submit:    //提交
                                if (!userPsd.equals(confirPsd)){
                                    Toast.makeText(RegActivity.this,"两次密码不一致，注册失败",Toast.LENGTH_SHORT).show();
                                }
                                else {
                                Intent intent=new Intent(RegActivity.this, ShowActivity.class);
                                Bundle bundle=new Bundle();
                                bundle.putString("name",userName);
                                bundle.putString("password", userPsd);
                                bundle.putString("email", userEmail);
                                bundle.putString("phone", userPhone);
                                intent.putExtras(bundle);
                                startActivity(intent);
                                }
                                break;
                            case R.id.btn_return://返回
                                Intent intent1=new Intent();
                                intent1.putExtra("name", userName);
                                intent1.putExtra("password", userPsd);
                                setResult(102, intent1);
                                finish();
                                break;
                        }
                    }
                }
```

（5）在 ShowActivity 中处理传递过来的数据。

在 onResume()方法中通过调用 getIntent().getExtras()获取 Bundle，得到 Bundle 中的注册信息并显示出来。

```java
            public class ShowActivity extends AppCompatActivity {
                private TextView textView;
                private Button button;
                @Override
                protected void onCreate(Bundle savedInstanceState) {
                    super.onCreate(savedInstanceState);
                    setContentView(R.layout.activity_show);
                    setTitle("注册信息");
                    textView=findViewById(R.id.tv_information);
                    button=findViewById(R.id.button);
                    button.setOnClickListener(new View.OnClickListener() {
                        @Override
                        public void onClick(View v) {
                            finish();
                        }
                    });
                }
                @Override
                protected void onResume() {
                    super.onResume();
                    Intent intent =getIntent();
```

```
            Bundle bundle=intent.getExtras();
            StringBuilder info=new StringBuilder();
            info.append("姓名："+bundle.get("name")+"\n") ;
            info.append("密码："+bundle.get("password")+"\n") ;
            info.append("邮箱："+bundle.get("email")+"\n") ;
            info.append("手机号："+bundle.get("phone")+"\n") ;
            textView.setText(info);
        }
    }
```

（6）在 LoginActivity 中处理 RegActivity 传回的数据。

当用户单击注册界面上的"返回登录界面"按钮时，LoginActivity 中的 onActivityResult() 方法会被调用，对 RegActivity 返回的数据进行处理。

```
            protected void onActivityResult(int requestCode, int resultCode, @Nullable Intent data) {
                super.onActivityResult(requestCode, resultCode, data);
                if (requestCode==101)
                    if (resultCode==102){
                        edName.setText(data.getStringExtra("name"));
                        edPswd.setText(data.getStringExtra("password"));
                    }
            }
```

4.6 使用 Fragment

Fragment（碎片）是一种可以嵌入在 Activity 中的 UI 片段，用来描述 Activity 的一部分布局。Fragment 具有生命周期和 UI 布局，并且能够响应事件，因此通常被封装成可重用的模块。一个 Activity 里可以有多个 Fragment，一个 Fragment 可以被多个 Activity 重用。

Fragment 能够同时兼顾手机和平板电脑的开发，可以灵活地为不同大小屏幕的设备创建 UI 界面。例如，当程序运行在一个大屏幕设备时可启动一个包含多个 Fragment 的 Activity，当程序运行在一个小屏幕设备时，可启动一个包含少量 Fragment 的 Activity。使用 Fragment 可以更加方便地在程序运行过程中动态地更新 Activity 的用户界面。在平板电脑上，Fragment 展现了很好的适应性和动态创建 UI 的能力。

4.6.1 Fragment 的生命周期

由于 Fragment 需要嵌入在 Activity 中使用，所以其生命周期和它所在的 Activity 是密切相关的。例如，当 Activity 暂停时，其中的所有 Fragment 也会暂停。当 Activity 被销毁时，所有 Fragment 也会被销毁。当 Activity 处于运行状态时，可以单独地对每一个 Fragment 进行操作，如添加或删除。当添加时，Fragment 处于启动状态；当删除时，Fragment 处于销毁状态。

Fragment 生命周期及其回调方法如图 4.21 所示。

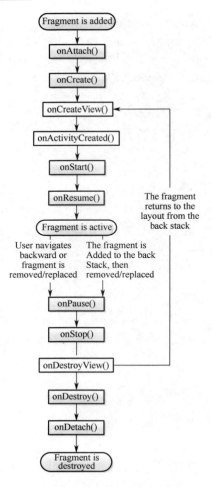

图 4.21　Fragment 的生命周期

Activity 直接影响其所包含的 Fragment 生命周期，对 Activity 生命周期中某个方法调用时，也会产生对 Fragment 相应方法的调用。例如，当 Activity 的 onPause()方法被调用时，其中所包含的所有 Fragment 的 onPause()方法都将被调用。

在生命周期中，Fragment 的回调方法比 Activity 多，主要用于与 Activity 进行交互，如 onAttach()、onCreateView()、onActivityCreated()、onDestroyView()和 onDetach()等方法。

当 Activity 进入运行状态时，允许添加和删除 Fragment。因此，只有当 Activity 处于 Resumed 状态时，Fragment 的生命周期才能独立运转，其他阶段都依赖于 Activity 的生命周期。

Fragment 生命周期的方法见表 4.4。

表 4.4　Fragment 生命周期的方法

方　　法	功　能　描　述
onAttach()	当 Fragment 和 Activity 相关联时调用
onCreate()	当 Fragment 被创建时调用
onCreateView()	当 Activity 获得 Fragment 的布局时调用，Fragment 在其中创建视图（加载布局）
onActivityCreated()	当 Fragment 所在的 Activity 启动完成时调用

（续表）

方　　法	功　能　描　述
onStart()	当 Fragment 在 UI 界面可见时调用
onResume()	当 Fragment 的 UI 界面可以与用户交互时调用
onPause()	当 Fragment 可见，但不可交互，Activity 转为 onPause 状态时调用
onStop()	当 Fragment 所依附的 Activity 转为 onStop 状态，或者 Activity 正在执行一个修改 Fragment 的操作而导致 Fragment 不再跟用户交互时调用
onDestroyView()	当 Fragment 清理 View 资源时调用，即移除 Fragment 中的视图
onDestroy()	当 Fragment 被销毁时调用
onDetach()	当 Fragment 被从 Activity 中删除时调用

4.6.2　创建 Fragment

创建 Fragment 的方法与 Activity 类似，可以在 Android Studio 环境中通过菜单命令来创建，也可以通过自定义 Fragment 子类的方法来创建。

在应用程序的包名处右击，并在弹出的菜单中执行"New"→"Fragment"→"Fragment(Blank)"菜单命令，打开"New Android Component"对话框，填写相关信息，然后单击"Finish"按钮，创建一个 Fragment 和其对应的布局文件，如图 4.22 所示。

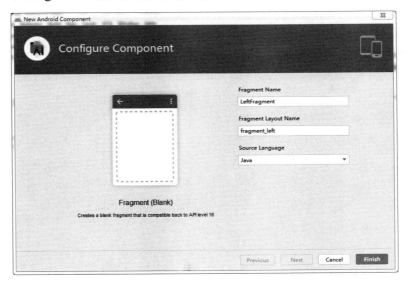

图 4.22　创建 Fragment

无论采用哪种方法创建的 Fragment，都必须继承 Fragment 类或其子类，实现 Fragment 中的回调。对于大部分 Fragment 而言，通常需要实现 onCreate()、onCreateView()和 onPause() 三种方法。系统在第一次绘制 Fragment 对应的 UI 时需要调用 onCreateView()方法，并在该方法中调用 inflater.inflate()方法加载 Fragment 的布局文件，返回加载的 View 对象。如果 Fragment 未提供 UI 则返回 null。

inflate(int resource, ViewGroup root, boolean attachToRoot)

其中，参数 resource 为 Fragment 需要加载的布局文件；参数 root 为加载 Fragment 的父

ViewGroup，即 onCreateView 传递进来的 container；参数 attachToRoot 为是否返回父 ViewGroup，取值应为 false，因为在 Fragment 内部实现中，会把该布局添加到 container 中，如果设为 true，则会重复做两次添加。

创建 Fragment 子类的代码如下：

```java
public class RightFragment extends Fragment {
    @Override
    public View onCreateView(LayoutInflater inflater, ViewGroup container,
                             Bundle savedInstanceState) {
        View view = inflater.inflate(R.layout.fragment_right, container, false);
        return view;
    }
}
```

上述代码中 fragment_right 为 RightFragment 的布局文件。

4.6.3 在 Activity 中添加 Fragment

Fragment 不能独立运行，必须嵌入 Activity 中使用。将 Fragment 添加到 Activity 中有以下两种方法。

1．在 Activity 的布局文件中添加 Fragment

在 Activity 的布局文件中通过<fragment>标签添加 Fragment，<fragment>标签与其他组件的标签类似，但必须指定 name 属性和一个唯一标识符，name 属性值为 Fragment 的全路径名称，标识符可以为 ID 或 tag。

例如，在 Activity 的布局文件 activity_main.xml 中添加 RightFragment 的代码如下：

```xml
<LinearLayout xmlns:android="http://schemas.android.com/apk/res/android"
    xmlns:tools="http://schemas.android.com/tools"
    android:layout_width="match_parent"
    android:layout_height="match_parent"
    android:orientation="horizontal"
    tools:context=".MainActivity">
    <fragment
        android:name="com.example.fragment.RightFragment"
        android:id="@+id/right_fg"
        android:layout_width="match_parent"
        android:layout_height="match_parent"/>
</LinearLayout>
```

当创建 Activity 的布局文件时，系统会实例化每一个 Fragment，并且调用其 onCreateView() 方法来获得相应 Fragment 的布局，并将返回值插入 Fragment 标签所在的位置。

在 Activity 的布局文件中添加 Fragment，会使 Fragment 及其视图与 Activity 绑定在一起，无法灵活地切换 Fragment。在实际开发中，多采用在 Activity 运行时将 Fragment 添加到 Activity 布局容器的方法。

2．在 Activity 的代码中添加 Fragment

当 Activity 添加时，Fragment 需要获取一个 FragmentManager 的实例，然后调用 FragmentTransaction 实例 add()、replace()和 remove()的方法进行 Fragment 的添加、替换和删

除操作,最后通过 commit()方法提交事务。

将 Fragment 动态添加到 Activity 的布局中,需要有一个存放 Fragment 的容器,一般是 FrameLayout。

【例 4.6】Fragment 的创建和加载。

本例演示在一个 Activity 中展示左右两个 Fragment,左边的 LeftFragment 添加在主界面 MainActivity 的布局文件 activity_main.xml 中,右边的 RightFragment 和 ReplaceFragment 添加在代码中,这两个 Fragment 可以动态切换。

(1) 创建 LeftFragment。

创建一个应用程序,包名为 com.example.fragment。在应用程序中创建 LeftFragment,对应的布局文件为 left_fragment.xml,并在该布局中添加两个 Button 控件。left_fragment.xml 文件的代码如下:

```xml
<LinearLayout xmlns:android="http://schemas.android.com/apk/res/android"
    android:layout_width="match_parent"
    android:layout_height="match_parent"
    android:orientation="vertical" >
    <Button
        android:id="@+id/button"
        android:layout_marginTop="30dp"
        android:layout_width="wrap_content"
        android:layout_height="wrap_content"
        android:textSize="30dp"
        android:text="按钮 1"/>
    <Button
        android:id="@+id/button_another"
        android:layout_marginTop="20dp"
        android:layout_width="wrap_content"
        android:layout_height="wrap_content"
        android:textSize="30dp"
        android:text="按钮 2" />
</LinearLayout>
```

(2) 创建 RightFragment。

在应用程序中创建 RightFragment,对应的布局文件为 right_fragment.xml,并在该布局中添加一个 TextView 控件。right_fragment.xml 文件的代码如下:

```xml
<LinearLayout xmlns:android="http://schemas.android.com/apk/res/android"
    android:layout_width="match_parent"
    android:layout_height="match_parent"
    android:gravity="center"
    android:orientation="vertical" >
    <TextView
        android:layout_width="wrap_content"
        android:layout_height="wrap_content"
        android:textSize="20sp"
        android:text="This is right fragment"
        />
</LinearLayout>
```

（3）创建 ReplaceFragment。

在应用程序中创建 ReplaceFragment，对应的布局文件为 replace_fragmen.xml，并在该布局中添加一个 TextView 控件。replace_fragmen.xml 文件的代码如下：

```xml
<LinearLayout xmlns:android="http://schemas.android.com/apk/res/android"
    android:layout_width="match_parent"
    android:layout_height="match_parent"
    android:gravity="center"
    android:orientation="vertical" >
    <TextView
        android:layout_width="wrap_content"
        android:layout_height="wrap_content"
        android:textSize="20sp"
        android:text="This is another right fragment"
        />
</LinearLayout>
```

（4）主界面设计。

在 activity_main.xml 文件中添加一个 Fragment 和一个 FrameLayout 的布局。activity_main.xml 文件的代码如下：

```xml
<LinearLayout xmlns:android="http://schemas.android.com/apk/res/android"
    android:layout_width="match_parent"
    android:layout_height="match_parent">
    <!-- 静态加载 LeftFragment -->
    <fragment
        android:id="@+id/left_fragment"
        android:name="com.example.fragment.LeftFragment"
        android:layout_width="0dp"
        android:layout_height="match_parent"
        android:layout_weight="1" />
    <View        <!--分割线 -->
        android:layout_width="1dp"
        android:layout_height="match_parent"
        android:background="@color/colorPrimaryDark"/>
    <FrameLayout    <!--在这个容器中动态加载右边的 Fragment >
        android:id="@+id/right_layout"
        android:layout_width="0dp"
        android:layout_height="match_parent"
        android:layout_weight="3" >
    </FrameLayout>
</LinearLayout>
```

（5）实现 RightFragment 和 ReplaceFragment 的 onCreateView( )方法。

```java
public class RightFragment  extends Fragment {
    public View onCreateView(LayoutInflater inflater, @Nullable ViewGroup container, Bundle savedInstanceState) {
        View view=inflater.inflate(R.layout.right_fragment,container,false);
        return view;
    }
}
```

```java
public class ReplaceFragment extends Fragment {
    public View onCreateView(LayoutInflater inflater, @Nullable ViewGroup container, Bundle savedInstanceState) {
        View view=inflater.inflate(R.layout.replace_fragmen,container,false);
        return view;
    }
}
```

（6）编写 LeftFragment 的按钮事件处理代码。

单击"按钮1"按钮，动态加载 RightFragment；单击"按钮2"按钮，动态加载 ReplaceFragment，替换为 RightFragment。

```java
public class LeftFragment extends Fragment    implements View.OnClickListener{
    FragmentManager fragmentManager;
    public View onCreateView(LayoutInflater inflater, @Nullable ViewGroup container, Bundle savedInstanceState) {
        View view=inflater.inflate(R.layout.left_fragment,container,false);
        Button button1=view.findViewById(R.id.button);
        Button button2=view.findViewById(R.id.button_another);
        fragmentManager=getFragmentManager();
        button1.setOnClickListener(this);
        button2.setOnClickListener(this);
        return view;
    }
    @Override
    public void onClick(View v) {
        switch (v.getId()){
            case R.id.button:
                //按钮1，动态加载 RightFragment
                RightFragment fragment=new RightFragment();
                FragmentTransaction transaction=fragmentManager.beginTransaction();
                transaction.replace(R.id.right_layout,fragment);
                transaction.commit();
                break;
            case  R.id.button_another:
                //按钮2，动态加载 ReplaceFragment, 替换为 RightFragment
                ReplaceFragment replaceFragment=new ReplaceFragment();
                FragmentTransaction anothertransaction=fragmentManager.beginTransaction();
                anothertransaction.replace(R.id.right_layout,replaceFragment);
                anothertransaction.commit();
                break;
        }
    }
}
```

运行程序，单击图 4.23 所示的主界面上的"按钮1"按钮，动态加载 RightFragment，如图 4.24 所示。单击"按钮2"按钮，动态加载 ReplaceFragment，替换为 RightFragment，如图 4.25 所示。

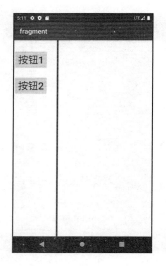 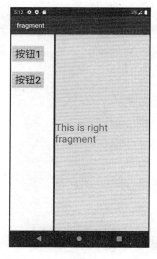 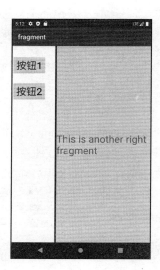

图 4.23　主界面　　　图 4.24　动态加载 RightFragment　　　图 4.25　动态加载 ReplaceFragment

4.6.4　Activity 与 Fragment 的通信

Fragment 与 Activity 之间的通信包括两个方面：Activity 向 Fragment 传递数据；Fragment 向 Activity 传递数据。Fragment 与 Activity 的通信可以通过回调接口实现，如图 4.26 所示。

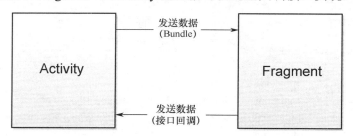

图 4.26　Fragment 与 Activity 通信示意

1．Activity 向 Fragment 传递数据

Activity 向 Fragment 传递数据最简单的方法是在 Activity 中创建 Bundle 数据包，并调用 setArguments(Bundle bundle)将 Bundle 数据包传递给 Fragment。

```
FragmentManager fragmentManager = getFragmentManager();
FragmentTransaction fragmentTransaction = fragmentManager.beginTransaction();
final mFragment fragment = new mFragment();
Bundle bundle = new Bundle();
bundle.putString("message", "I love Android");
fragment.setArguments(bundle);
fragmentTransaction.add(R.id.fragment_container, fragment);
fragmentTransaction.commit();
```

在 Fragment 中，通过 getArgments()获取从 Activity 传过来的数据。

```
Bundle bundle = this.getArguments();
```

```
            String message = bundle.getString("message");
```
当 Fragment 创建后，需要用到下面两种方法获取 Activity 所包含的 Fragment 实例。
```
            getSupportFragmentManager().findFragmentById(R.id.fragment_container);
            getSupportFragmentManager().findFragmentByTag("tag");
```
在布局文件中使用<fragment>标签添加 Fragment 时，可以为 Fragment 指定 android:id 或 android:tag 属性，这两个属性都可用于标识 Fragment。

2．Fragment 向 Activity 传递数据

Fragment 向 Activity 传递数据通常借助于回调接口。在 Fragment 中定义一个回调接口，与之通信的 Activity 实现该接口，Fragment 通过调用该回调接口方法与 Activity 或其他 Fragment 共享信息。

Android 引入 Fragment 的初衷是为了适应大屏幕的平板电脑，在一个 Activity 中展示两个 Fragment，并实现 Activity 与 Fragment 的通信，模拟平板电脑中新闻的显示。

【例 4.7】使用 Fragment 模拟平板电脑中新闻的显示。

本例中应用程序的 MainActivity 中包含两个 Fragment：TitleFragment 继承自 ListFragment 类，用于显示新闻标题列表；ContentFragmen 用于显示标题对应的新闻内容。在 TitleFragment 中，用户单击某个新闻标题时通知 MainActivity，然后 MainActivity 再通知 ContentFragmen，此时 ContentFragmen 就会显示该标题对应的新闻内容。在 TitleFragment 中定义 OnNewsSelectedListener 接口，然后在 MainActivity 中实现该接口，并重写 onNewsSelected() 方法以通知 ContentFragmen 来自于 TitleFragment 的事件。当 Fragment 添加到 MainActivity 中时，会调用 Fragment 的 onAttach() 方法，在该方法中检查 MainActivity 是否实现了 OnNewsSelectedListener 接口，并对传入的 MainActivity 实例强制转换为接口变量。

由于 TitleFragment 继承自 ListFragment 类，所以每次选中列表项时，都会调用 TitleFragment 的 onListItemClick()方法，在 onListItemClick()方法中调用 onNewsSelected()方法，实现与 Activity 的共享事件。

（1）创建 TitleFragment 的布局文件 title.xml。
```
            <?xml version="1.0" encoding="utf-8"?>
            <LinearLayout xmlns:android="http://schemas.android.com/apk/res/android"
                android:layout_width="match_parent"
                android:layout_height="match_parent">
                <ListView
                    android:id="@id/android:list"
                    android:layout_width="match_parent"
                    android:layout_height="match_parent"/>
            </LinearLayout>
```
（2）创建 ContentFragmen 的布局文件 content.xml。

在该布局文件中添加两个 TextView 控件，上边的 TextView 用于显示新闻标题，下边的 TextView 用于显示新闻内容。
```
            <?xml version="1.0" encoding="utf-8"?>
            <LinearLayout xmlns:android="http://schemas.android.com/apk/res/android"
                android:orientation="vertical"
```

```xml
        android:layout_width="match_parent"
        android:layout_height="match_parent">
    <TextView
        style="?android:attr/textAppearanceLarge"
        android:id="@+id/new_title"
        android:padding="16dp"
        android:layout_width="match_parent"
        android:layout_height="wrap_content"/>
    <TextView
        style="?android:attr/textAppearanceMedium"
        android:id="@+id/new_content"
        android:padding="5dp"
        android:layout_width="match_parent"
        android:layout_height="match_parent"/>
</LinearLayout>
```

（3）在布局文件 activity_main.xml 中添加 Fragment。

在 MainActivity 的布局文件 activity_main.xml 中，左部添加一个<fragment>元素，表示显示新闻标题列表的 TitleFragment，右部添加一个 FrameLayout 容器，用于放置展示新闻内容的 ContentFragment。

```xml
<?xml version="1.0" encoding="utf-8"?>
<LinearLayout xmlns:android="http://schemas.android.com/apk/res/android"
    xmlns:tools="http://schemas.android.com/tools"
    android:layout_width="match_parent"
    android:layout_height="match_parent"
    android:orientation="horizontal"
    android:divider="?android:attr/dividerHorizontal"
    android:showDividers="middle"
    tools:context=".MainActivity">
    <fragment
        android:name="com.example.newsfragment.TitleFragment"
        android:id="@+id/title"
        android:layout_weight="1.5"
        android:layout_width="0dp"
        android:layout_height="wrap_content"/>
    <FrameLayout
        android:id="@+id/content_container"
        android:layout_weight="2.5"
        android:layout_width="0dp"
        android:layout_height="match_parent"/>
</LinearLayout>
```

（4）创建 TitleFragment。

创建 TitleFragment 类，继承于 ListFragment 类。在 TitleFragment 类中定义一个回调接口 OnNewsSelectedListener，TitleFragment 所在的 MainActivity 需要实现该接口。

```java
public class TitleFragment extends ListFragment {
    OnNewsSelectedListener mListener;
    //定义回调接口
    public interface  OnNewsSelectedListener{
            public void onNewsSelected(int id);
        }
    @Override
    public View onCreateView(LayoutInflater inflater, ViewGroup container,  Bundle savedInstanceState) {
        View view=inflater.inflate(R.layout.title,container,false);
        MainActivity activity=(MainActivity) getActivity();
        String titles[]=activity.getNewsTitles();
        //创建适配器
        ArrayAdapter<String>adapter=new ArrayAdapter<String>(getActivity(),
                android.R.layout.simple_expandable_list_item_1,titles);
        //绑定适配器时，必须通过 ListFragment.setListAdapter 接口
        setListAdapter(adapter);
        return view;
    }
    @Override
        public void onAttach( Context context) {
            super.onAttach(context);
//如果 Activity 没有实现 OnNewsSelectedListener 接口，则抛出异常
    try{
            mListener=(OnNewsSelectedListener)context;
    }catch (ClassCastException e){
            throw new ClassCastException(context.toString()+"必须继承
                OnNewsSelectedListener 接口");
        }
    }
    @Override
//重写 onListItemClick()方法，当用户单击某列表项时回调该方法
public void onListItemClick( ListView l,    View v, int position, long id) {
//调用接口中的 onNewsSelected()方法
mListener.onNewsSelected(position);
}
@Override
public void onDetach() {
        super.onDetach();
        mListener=null;
    }
}
```

（5）创建 ContentFragment。

创建 ContentFragment,继承于 Fragment 类,实现 onCreateView()方法,并编写一个 reflush() 方法,用于刷新 ContentFragment 上控件的内容。

```java
public class ContentFragment extends Fragment {
    private TextView tv_title;
    private TextView tv_content
    private View rootView;
    @Override
    public void onCreate(Bundle savedInstanceState) {
        super.onCreate(savedInstanceState);
    }
    @Override
    public View onCreateView(LayoutInflater inflater,ViewGroup container, Bundle
                    savedInstanceState) {
        rootView=inflater.inflate(R.layout.content,container,false);
        tv_title=(TextView)rootView.findViewById(R.id.new_title);
        tv_content=(TextView)rootView.findViewById(R.id.new_content);
        return rootView;
    }
    // 刷新 ContentFragment 上控件的内容
    public void reflush(String newsTitle,String newsContent){
            tv_title.setText(newsTitle);
            tv_content.setText(newsContent);
    }
}
```

（6）编写 MainActivity。

MainActivity 需要实现 TitleFragment 类中定义的回调接口 OnNewsSelectedListener，重写其中的 onNewsSelected(int id)方法。

```java
public class MainActivity extends AppCompatActivity implements TitleFragment.OnNewsSelectedListener {
    ContentFragment fragment;
    //新闻标题
    private String newsTitles[]={"服务员制止餐厅浪费",
            "戍边 20 多年 13 次与死神擦肩",
            "多地中小学开学时间确定"
    };
    //新闻内容
    private String newsContents[]={"北京日报客户端 2 月 21 日消息，这几天，北京丰台区某餐厅服务员因制止顾客餐饮浪费反被指责，引起网上持续热议。网友们纷纷表示，要以餐饮浪费为耻，为光盘行动撑腰。",
            "卫国戍边英雄团团长祁发宝 20 多年戍边生涯中曾 13 次与死神擦肩。他说：'不是所有人都能理解我的选择，但我无怨无悔！'",
            "随着春节假期的结束，全国各地的大中小学陆续公布了开学时间，根据教育部的要求，全国各高校可自主调整开学时间。"
    };
    public String[]getNewsTitles(){
        return newsTitles;
    }
    public String[]getNewsContents(){
        return newsContents;
    }
    @Override
    protected void onCreate(Bundle savedInstanceState) {
```

```
            super.onCreate(savedInstanceState);
            setContentView(R.layout.activity_main);
            fragment=new ContentFragment();
            FragmentManager fragmentManager=getSupportFragmentManager();
            FragmentTransaction transaction=fragmentManager.beginTransaction();
            transaction.add(R.id.content_container,fragment);
            transaction.commit();
        }
        @Override
        // 重写接口中的方法
        public void onNewsSelected(int id) {
            fragment.reflush(newsTitles[id],newsContents[id]);
        }
    }
```

运行结果如图 4.27 所示。

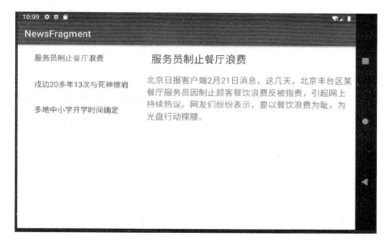

图 4.27　运行结果

习题 4

1. 编写程序，通过 LogCat 工具观察 Activity 生命周期方法的回调。
2. 编写程序，使用隐式 Intent 启动浏览器打开百度官网首页，选中联系人，启动拨号程序，通过内置电子邮件工具发送邮件。
3. 完成习题 2 中考试系统的登录、用户注册和修改密码功能。
（1）在登录界面输入正确的账号和密码，单击"登录"按钮切换到考试系统界面，在该界面显示"×××（账号），您好！欢迎进入考试系统！"（见图 4.28），单击该界面上的"退出"按钮退出应用程序。
（2）单击"注册"按钮（见图 4.29），切换到注册信息显示界面，在该界面上显示用户信息（见图 4.30）。
（3）单击用户信息界面的"修改密码"按钮，切换到修改密码界面（见图 4.31），输入新密码，单击"确定"按钮，返回到用户信息界面，并显示新密码（见图 4.32）。

图 4.28　考试系统界面　　图 4.29　用户注册界面　　图 4.30　用户信息

图 4.31　修改密码界面　　图 4.32　显示新密码

第 5 章 Android 数据存储

在计算机程序的运行过程中,会获取或生成各种各样的数据,并通过各种方式在系统中保存下来便于后续的查询和使用。Android 系统也不例外,几乎 Android 系统的每个 App 都需要存储数据以保存用户或程序的设置信息、状态信息或其他程序运行需要的数据,因此需要有效的手段及方式来对以上信息加以存储。Android 系统提供了多种数据存储方式,可以有效地应对以上需求。本章介绍 SharedPreferences、内部及外部文件存储、SQLite 数据库等多种数据存储方式,以及各种数据存储方式适用的场景及相关注意事项。

5.1 SharedPreferences 数据存储

SharedPreferences 是 Android 提供的一个轻量级数据存储类,一般用于存储应用程序的设置数据或少量其他数据。SharedPreferences 通过键-值对的形式来存储数据,其数据存储在 XML 格式文件中,存放在/data/data/<package name>/shared_prefs 目录下。

下面介绍使用 SharedPreferences 实现数据保存、读取和修改的方法。

5.1.1 使用 SharedPreferences 存储数据

在 SharedPreferences 中存储数据(键-值对),需要使用 SharedPreferences.Editor 类,通过调用 SharedPreferences 的 edit()方法获取 Editor 对象,然后使用 Editor 对象的 putXXX( )方法来存储数据。SharedPreferences.Editor 接口的常用方法如表 5.1 所示。

表 5.1 SharedPreferences.Editor 接口的常用方法

方 法 名	说　　明
putBoolean(String key, boolean value)	保存一个 Boolean 类型数据
putFloat(String key, float value)	保存一个 Float 类型数据
putInt(String key, int value)	保存一个 Int 类型数据
putLong(String key, long value)	保存一个 Long 类型数据
putString(String key, String value)	保存一个 String 类型数据
putStringSet(String key, Set<String> values)	保存字符串集合类型的数据
remove(String key)	删除键名 key 所对应的数据值
clear()	清除所有数据值
commit()	保存数据

1. 保存数据

使用 SharedPreferences 保存数据需要以下四个步骤。

（1）获取 SharedPreferences 对象。

每个应用程序都有一个 SharedPreferences 对象，通过 Context 对象的 getSharedPreferences (String name, int mode)方法获取属于当前程序，并且名称为 name 的 SharedPreferences 对象。

其中，第一个参数 name 是指 SharedPreferences 对象对应的名称，也是其在文件系统中的文件名；第二个参数 mode 为操作模式，取值可以是以下三种：

- Context.MODE_PRIVATE：该文件只能被当前的应用程序访问。
- Context.MODE_WORLD_READABLE：该文件除当前程序访问外，还可以被其他应用程序读取。
- Context.MODE_WORLD_WRITEABLE：该文件除当前程序访问外，还可以被其他应用程序读取和写入。

从 Android 4.2 开始不再推荐使用 MODE_WORLD_READABLE 和 MODE_WORLD_WRITEABLE 这两种模式。

获取 SharedPreferences 对象的代码如下：

```
SharedPreferences prefs =getSharedPreferences("data"，MODE_PRIVATE);
```

（2）获取 SharedPreferences.Editor 编辑对象。

使用 SharedPreferences 对象的 edit()方法，获取 SharedPreferences.Editor 编辑对象。

```
SharedPreferences.Editor editor= prefs.edit();
```

（3）存储数据。

使用 SharedPreferences.Editor 对象的 putXXX( )方法来存储数据。

```
editor.putString("name", "张华");   //存入 String 类型数据
editor.putInt("age",8);            //存入 Int 类型数据
```

（4）提交数据。

调用 putXXX()方法之后，一定要调用 SharedPreferences.Editor 编辑对象的 apply()或 commit()方法将数据提交到 XML 文件中。

```
editor.commit();或者
editor. apply();
```

这两种方法的区别：commit()方法是同步提交，且有布尔类型的返回值可以判断提交是否成功；apply()方法是异步提交，且没有返回值。在磁盘操作不频繁时，这两种方法并没有显著的性能差异。

2. 修改数据

Editor 类并没有专门修改数据的方法，如果要修改数据，只需要重新调用 putXXX()方法，用一个新的值去覆盖同名的键值即可。

3. 删除数据

可以调用 remove()方法删除 SharedPreferences 中的单个键值对：

```
editor.remove(String key)
```

若要清空 SharedPreferences 中保存的全部键值对，可使用 clear()方法：

```
editpr.clear()
```

完成以上调用后，同样需要调用 Editor 对象的 apply()或 commit()方法提交，才能使删除操作生效。

5.1.2 使用 SharedPreferences 读取数据

首先需要获取 SharedPreferences 对象，然后使用该对象的 getXXX()方法通过键的名称取出相应值。SharedPreferences 类的常用方法如表 5.2 所示。

表 5.2 SharedPreferences 类的常用方法

方 法 名	作 用
boolean getBoolean(String key, boolean defValue)	从 SharedPreferences 对象中提取布尔类型的值
float getFloat(String key, float defValue)	从 SharedPreferences 对象中提取浮点数类型的值
int getInt(String key, int defValue)	从 SharedPreferences 对象中提取整数类型的值
long getLong(String key, long defValue)	从 SharedPreferences 对象中提取长整数类型的值
String getString(String key, String defValue)	从 SharedPreferences 对象中提取字符串类型的值
Set\<String\> getStringSet(String key, Set\<String\> defValues)	从 SharedPreferences 对象中提取字符串集合类型的值

getXXX()方法的第一个参数是需要取出值对应的键名，第二个参数是默认值，即当 SharedPreferences 没有找到这个键值对时，getXXX()方法返回的默认值。

5.1.3 SharedPreferences 使用示例

【例 5.1】用 SharedPreferences 存储用户的注册信息。

修改【例 4.5】登录程序的登录、注册功能，将用户的注册信息存储在 SharedPreferences 中，登录时从 SharedPreferences 中读取用户信息进行验证。

具体要求：在登录界面单击"注册"按钮时，跳转到注册界面，并单击注册界面上的"提交"按钮，将注册信息保存到 SharedPreferences 中。单击登录界面的"登录"按钮时，从 SharedPreferences 中读取用户信息进行身份验证，登录成功后跳转到用户界面，在该界面显示当前用户名。

（1）修改注册功能。

修改注册界面上"提交"按钮的单击事件处理代码，将用户填写的注册信息保存到 SharedPreferences 中。将 RegActivity 中"提交"按钮的事件处理代码修改如下：

```
case R.id.btn_submit:    //提交
    if (!userPsd.equals(confirPsd)){
        Toast.makeText(RegActivity.this,"两次密码不一致，注册失败",Toast.LENGTH_SHORT).show();
    }
    else {
        SharedPreferences sp = getSharedPreferences("userinfo", MODE_PRIVATE);
        SharedPreferences.Editor editor = sp.edit();
        editor.putString("username", userName);
        editor.putString("password", userPsd);
        editor.putString("email", userEmail);
        editor.putString("phone", userPhone);
```

```
                    editor.commit();
                    Toast.makeText(RegActivity.this,"注册成功",Toast.LENGTH_SHORT).show();
                }
                break;
```

（2）修改登录功能。

修改登录界面上"登录"按钮的事件处理代码，从 SharedPreferences 中读取用户信息进行正确性验证。将 LoginActivity 中"登录"按钮的事件处理代码修改如下：

```
            case R.id.btn_login:
                String userName=edName.getText().toString();
                String password=edPswd.getText().toString();
                if (userName.isEmpty()) {
                    Toast.makeText(LoginActivity.this, "请输入用户名", Toast.LENGTH_LONG).show();
                        return;
                    }
                if (password.isEmpty()){
                    Toast.makeText(LoginActivity.this, "请输入密码", Toast.LENGTH_LONG).show();
                        return;
                    }
                //获取 SharedPreferences 文件
                SharedPreferences sp = getSharedPreferences("userinfo",MODE_PRIVATE);
                //从 SharedPreferences 文件中获取用户名和密码
                String name=sp.getString("username", "");
                String pswd=sp.getString("password", "");
                //用户名和密码是否一致，如一致则跳转到用户界面
                if (name.equals(userName) && pswd.equals(password)) {
                    Intent intent = new Intent(LoginActivity.this, UserActivity.class);
                    intent.putExtra("name", userName);    //将 userName 封装到 Intent 中
                    startActivity(intent);     //启动 UserActivity
                }
                else {
                    Toast.makeText(LoginActivity.this,"用户名或密码错误", Toast.LENGTH_LONG).show();
                    }
                break;
```

5.1.4 SharedPreferences 使用注意事项

使用 SharedPreferences 存储数据是 Android App 实现数据存储的便利方式，通常用来保存应用的一些设置信息和少量的常用数据，提供了 String、StringSet、Int、Long、Float、Boolean 六种类型数据的存储功能。另外，使用 SharedPreferences 存储数据就是把数据存储到当前应用程序的 prefs 目录下，文件名为 SharedPreferences 的 xml 文件，具体的路径为/data/ data/应用程序包名/shared_prefs。

（1）获取 SharedPreferences 对象的三种方法。

① Context 类中的 getSharedPreferences(arg1, arg2)方法，其中，arg1 代表文件名；arg2 代表操作模式（默认为 MODE_PRIVATE）。

② Activity 类中的 getPreferences(arg1)方法，其中 arg1 的操作模式同①一样，文件名默认为当前活动的类名，这种方式获取的 SharedPreferences 对象只能在当前 Activity 中使用。

③ PreferenceManager 类中的 getDefaultSharedPreferences(arg1)方法，其中 arg1 为 Context，文件名为当前应用程序的包名。

（2）卸载软件时会把 SharedPreferences 生成的文件删除。在不卸载软件重装时，其文件插入是叠加式的，即当遇到 key 名称与类型都一样时，会覆盖其值；如果其 key 值一样，但类型不同，则将其插入。

（3）不要存储较大的数据对象到 SharedPreferences 对象中。

SharedPreferences 对象只适合少量数据的存储，所以单个 SharedPreferences 对象内存储的键值对数量不能太多。

（4）尽量减少提交的次数，以提高数据操作的效率，如可以在插入或修改多项数据后统一提交。

（5）如果不需要返回值，则尽量使用 apply()方法提交操作。

5.2 Android 权限管理

在默认情况下，Android 应用不拥有任何权限，由于每个 Android 应用都运行在进程沙盒中，如果应用需要获取沙盒之外的资源和数据，则必须申请其所需的权限。

5.2.1 权限机制

在 Android 6.0 之前，权限的使用相当简单，只需要在 AndroidManifest.xml 文件中声明即可。例如，一个需要监控接收的 SMS 消息的应用只需指定：

`<uses-permission android:name="android.permission.RECEIVE_SMS" />`

为了更好地保护用户隐私，谷歌公司在 Android 6.0 时提出了新的权限管理机制。将系统权限分成正常权限和危险权限。正常权限不会直接给用户隐私权带来风险，而申请了危险权限的应用程序，可能涉及用户隐私信息的数据或资源，也可能对用户存储的数据或其他应用的操作产生影响。

正常权限的授权方式跟 Android 6.0 之前的版本一样，只需要在 AndroidManifest 中声明，系统就会自动授权。而危险权限除需要在 AndroidManifest 文件中声明外，还需要在使用时给用户明确授予。

为了方便管理某一类特定功能的权限，Android 系统把危险权限按照功能分了九组不同的集合，这个集合就是权限组，如表 5.3 所示。权限组内包含一项或多项功能和特性相关的权限，所有的危险权限都属于各自的权限组。

表 5.3 权限组

Permission Group	Permission
CALENDAR	android.permission.READ_CALENDAR android.permission.WRITE_CALENDAR
CAMERA	android.permission.CAMERA
CONTACTS	android.permission.READ_CONTACTS android.permission.WRITE_CONTACTS android.permission.GET_ACCOUNTS

（续表）

Permission Group	Permission
LOCATION	android.permission.ACCESS_FINE_LOCATION
	android.permission.ACCESS_COARSE_LOCATION
MICROPHONE	android.permission.RECORD_AUDIO
PHONE	android.permission.READ_PHONE_STATE
	android.permission.READ_PHONE_NUMBERS
	android.permission.ALL_PHONE
	android.permission.READ_CALL_LOG
	android.permission.WRITE_CALL_LOG
	android.permission.USE_SIP
	android.permission.PROCESS_OUTGOING_CALLS
	android.permission.ANSWER_PHONE_CALLS
	com.android.voicemail.permission.ADD_VOICEMAIL
SENSORS	android.permission.BODY_SENSORS
SMS	android.permission.SEND_SMS
	android.permission.RECEIVE_SMS
	android.permission.READ_SMS
	android.permission.RECEIVE_WAP_PUSH
	android.permission.RECEIVE_MMS
STORAGE	android.permission.READ_EXTERNAL_STORAGE
	android.permission.WRITE_EXTERNAL_STORAGE

应用程序在向用户申请危险权限时，系统会弹出对话框，描述应用程序要使用的权限组，这时用户如果同意授权，系统就会将属于同一权限组，并且在清单中注册的其他权限也一起授予应用程序。但是在 Android 8.0 及以后版本中，系统只会授予应用程序明确请求的权限。然而，一旦用户为应用程序授予了某个权限，所有后续对该权限组中权限的请求就会被自动批准。

例如，应用程序在 AndroidManifest 中声明 READ_EXTERNAL_STORAGE 权限和 WRITE_EXTERNAL_STORAGE 权限，应用程序动态请求 READ_EXTERNAL_STORAGE 权限，并且用户授予了该权限。在 Android 8.0 之前，系统会自动授予 WRITE_EXTERNAL_STORAGE 权限，因为该权限也属于 STORAGE 权限组，并且也在清单中声明过。但是在 Android 8.0 及以后版本中，系统只会授予 READ_EXTERNAL_STORAGE 权限，不过该应用程序后续再申请 WRITE_EXTERNAL_STORAGE 时，系统会立即授予该权限，并且不会提示用户。

5.2.2 运行时权限申请

运行时申请权限需要用到以下三种方法。

（1）public static int checkSelfPermission (Context context, String permission)。

该方法用于检查应用程序是否具有参数 permission 指定的权限，返回值为常量 PackageManager.PERMISSION_GRANTED 或 PackageManager.PERMISSION_DENIED。

（2）public static void requestPermissions (Activity activity, String[] permissions, int requestCode)。

该方法用于申请权限，参数 activity 代表当前 Activity 实例，其中，permissions 是申请的权限列表，requestCode 为请求码，回调时会用到。

（3）public abstract void onRequestPermissionsResult (int requestCode, String[] permissions, int[] grantResults)。

该方法用于处理请求权限的响应，当用户对请求权限做出响应之后，系统会回调该方法。其中，参数 requestCode 与申请权限时的 requestCode 对应，permissions 为权限列表，grantResults 为权限列表对应的返回值，用于判断 permissions 里面的每个权限是否申请成功。

Android 运行时权限的使用，主要分为以下四个步骤。

① 声明权限。

在 AndroidManifest.xml 文件中声明需要的权限。

② 检查是否拥有权限。

当应用程序需要使用某一个权限时，可使用 checkSelfPermission()方法检查是否已经授予该权限。

③ 申请权限。

如果 checkSelfPermission()方法的返回值为 PERMISSION_GRANTED，则应用程序继续后续操作。如果该方法的返回值为 PERMISSION_DENIED，则通过调用 requestPermissions()方法来请求所需的权限。

④ 处理权限回调。

调用 requestPermissions()方法申请授权时，系统将向用户显示一个对话框，当用户响应时，系统将调用 onRequstPermissionsResult()方法，向其传递用户的响应。在 Activit 中重写 onRequestPermissionsResult()方法，来处理权限申请结果。

【例 5.2】申请相机权限。

本例演示如何动态申请相机权限。Android 应用中如果要调用系统相机，则需要 CAMERA 权限。

（1）声明权限。

创建一个应用程序，在清单文件 AndroidManifest.xml 中，声明如下权限：

```
<uses-permission android:name="android.permission.CAMERA"/>
```

（2）主界面设计。

在 activity_main.xml 布局文件中添加一个 Button 控件。

（3）申请权限。

在 MainActivity 中编写 requestPermission()方法用来申请权限。首先，使用 checkSelfPermission()方法检查是否有打开相机的权限。如果已授权，则在 LogCat 窗口中输出相应的信息；如果未授权，则使用 requestPermissions()申请相机权限。然后，在按钮的单击事件中调用 requestPermission()方法。最后，重写 MainActivity 的 onRequestPermissionsResult()方法，在 LogCat 窗口中输出权限申请结果，其代码如下：

```
public class MainActivity extends AppCompatActivity {
    private Button button;
    private static final  int  REQUEST_CAMERA=1;   //返回码
    @Override
    protected void onCreate(Bundle savedInstanceState) {
        super.onCreate(savedInstanceState);
```

```java
        setContentView(R.layout.activity_main);

        button=findViewById(R.id.button);
        button.setOnClickListener(new View.OnClickListener() {
            @Override
            public void onClick(View v) {
                requestPermission();
            }
        });
    }
    public void requestPermission() {
        //判断 SDK 版本，Android 6.0 以上需要动态申请权限
        if (Build.VERSION.SDK_INT >= Build.VERSION_CODES.M) {
            //判断是否授权
            if (ContextCompat.checkSelfPermission(this,
                    Manifest.permission.CAMERA) ==
                        PackageManager.PERMISSION_GRANTED) {
                //已授权
                    Log.i("msg", "已拥有相机权限");
            }else { //未授权，申请
                    ActivityCompat.requestPermissions(this,
                        new String[]{Manifest.permission.CAMERA},
                        REQUEST_CAMERA);
            }
        }
    }
    //处理权限申请结果
    @Override
    public void onRequestPermissionsResult(int requestCode, @NonNull String[] permissions,
      @NonNull int[] grantResults) {
        super.onRequestPermissionsResult(requestCode, permissions, grantResults);
        if (requestCode == REQUEST_CAMERA) {
            if (grantResults.length > 0 && grantResults[0] ==
                    PackageManager.PERMISSION_GRANTED) {
                Log.i("msg", "相机权限申请成功");
            } else {
                Log.i("msg", "用户拒绝授权");
            }
        }
    }
}
```

运行程序，主界面如图 5.1 所示，单击"打开相机"按钮，弹出如图 5.2 所示的对话框，选择"拒绝"选项，LogCat 窗口输出如图 5.3 所示的信息。再次单击"打开相机"按钮，又一次弹出如图 5.2 所示的对话框，选择"仅在使用该应用时允许"选项，LogCat 窗口输出如图 5.4 所示的信息。此时再次单击"打开相机"按钮，不会弹出对话框，LogCat 窗口输出如图 5.5 所示的信息。

20:30
PermissionDemo
打开相机

图 5.1　主界面

20:12
PermissionDemo
允许PermissionDemo拍摄照片和录制视频吗?
仅在使用该应用时允许
仅限这一次
拒绝

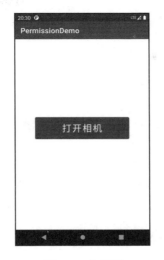

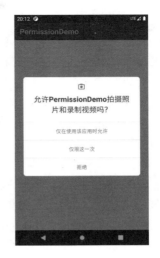

图 5.2　对话框

logcat
2021-05-08 20:14:12.521 I/msg: 用户拒绝授权

图 5.3　拒绝权限

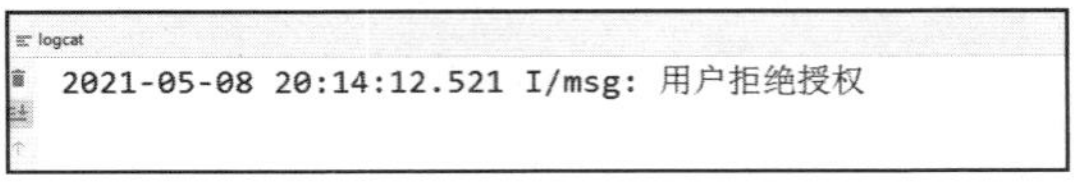

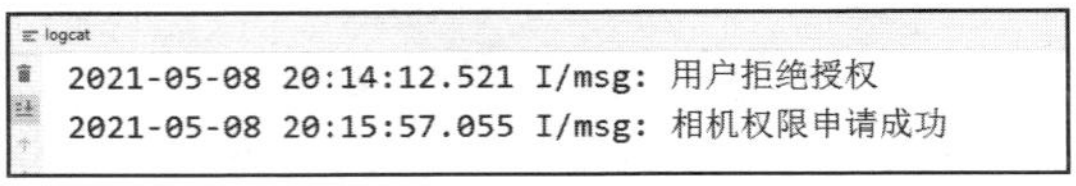

logcat
2021-05-08 20:14:12.521 I/msg: 用户拒绝授权
2021-05-08 20:15:57.055 I/msg: 相机权限申请成功

图 5.4　授权

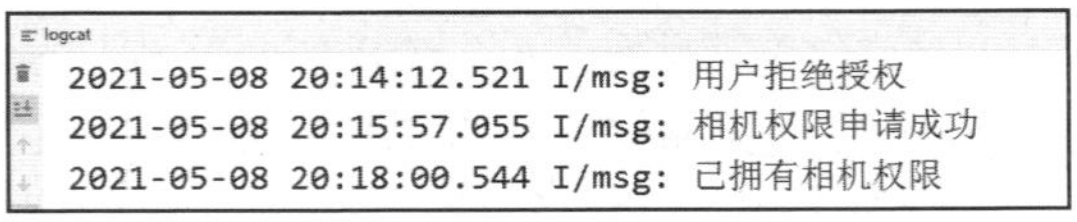

logcat
2021-05-08 20:14:12.521 I/msg: 用户拒绝授权
2021-05-08 20:15:57.055 I/msg: 相机权限申请成功
2021-05-08 20:18:00.544 I/msg: 已拥有相机权限

图 5.5　已拥有权限

5.3　数据的文件存储

5.3.1　Android 文件存储概述

和其他操作系统一样，在 Android 系统上，也可以把程序产生或运行所需的数据存储在文件中，Android 提供了相关的 API 接口，可方便开发人员执行文件的创建、读、写、删除等相关操作。

实质上，不管是 SharedPreferences 数据存储，还是数据库存储，本质上都是把数据存储在系统的某一个文件里。Android 系统上的文件存储，可以分为内部存储和外部存储。

内部存储指将应用程序的数据以文件方式存储在手机内置的存储区域中的应用程序专属空间，存放在这个区域的文件，只能由当前程序访问，其他应用程序无权进行操作。当创建的应用程序被卸载时，其内部存储的文件也随之被删除。

外部存储指把文件存放在外部存储区域（如 TF 卡或 SD 卡）内，存放在这个区域的文件，根据系统版本和放置位置的不同，其他程序会有不同的访问权限。

5.3.2 文件的内部存储

对于每个应用程序，系统都会在内部存储空间中提供相应目录用于存放文件。当前应用不需要任何系统权限即可读取和写入这些目录中的文件，而其他应用程序则无法访问，这使得内部存储空间非常适合存储应用程序的私有及敏感数据。存放持久性文件的目录一般为/data/data/package_name/files，存放缓存或临时文件的目录为/data/data/ package_name/cache，其中 package_name 指应用程序的包名。

访问内部文件有两种方式，一种是使用当前 Context 对象的 getFilesDir()方法，获取 File 对象所代表的目录，然后使用 Java IO API 实现数据访问；另一种是通过调用 openFileInput()方法和 openFileOutput()方法获取输入流和输出流，以实现文件的读/写操作。通常使用第二种方式进行内部文件的读/写操作，这跟 Java 中的 I/O 程序类似。

创建并读/写内部存储文件的步骤如下。

（1）通过 Context.openFileOutput(String name, int mode)方法打开文件并设定读/写方式。

该方法返回一个 FileOutputStream 对象用于写文件。其中，参数 name 为文件名，参数 mode 为文件操作模式，取值如下。

MODE_PRIVATE：默认访问方式，文件仅能被创建应用程序访问。

MODE_APPEND：若文件已经存在，则在文件末尾继续写入数据。

MODE_WORLD_READABLE：允许该文件被其他应用程序执行读操作。

MODE_WORLD_WRITEABLE：允许该文件被其他应用程序执行写操作。

从 Android 4.2.2 开始，常量 MODE_WORLD_READABLE 和 MODE_WORLD_WRITEABLE 已被弃用。从 Android 7 开始，使用这些常量将会引发 SecurityException。

（2）调用 FileOutputStream.write()方法写入数据。

（3）调用 FileOutputStream.close()方法关闭输出流，完成写操作。

以流的方式写文件代码如下：

```
String filename = "myfile";
String fileContents = "Hello world!";
try {
    FileOutputStream fos = context.openFileOutput(filename， Context.MODE_PRIVATE);
    fos.write(fileContents. getBytes() );
}
catch (IOException e) {
    e.printStackTrace();
}
```

以流的方式读取文件，须使用 openFileInput()，代码如下：

```
FileInputStream fis = context.openFileInput(filename);
InputStreamReader inputStreamReader =
        new InputStreamReader(fis, StandardCharsets.UTF_8);
StringBuilder stringBuilder = new StringBuilder();
try (BufferedReader reader = new BufferedReader(inputStreamReader)) {
    String line = reader.readLine();
    while (line != null) {
        stringBuilder.append(line).append('\n');
        line = reader.readLine();
```

```
            }
        } catch (IOException e) {
            e.printStackTrace();
        } finally {
            String contents = stringBuilder.toString();
        }
```

5.3.3 文件的外部存储

如果内部存储器的容量不足以存放应用程序的私有文件，则改为使用外部存储空间。Android 系统在外部存储空间中提供了专门的目录，应用程序可以在该目录下存放仅允许当前应用访问的文件。与内部存储类似，外部私有存储中的文件也分为持久性存储文件和缓存/临时文件，分别由一个目录与之相对应。

1. 外部存储中应用程序专属目录的访问

在 Android 4.4 或更高版本的系统中，应用程序无须请求任何与存储相关的权限，即可访问外部存储空间中的当前应用专属目录。卸载应用后，系统会移除在这些目录中存储的文件。

在 Android 9 或更低版本的系统里，只要其他应用具有相应的存储权限，都可以访问外部存储空间中的当前应用专属文件。为了确保应用数据的安全性，以 Android 10 及更高版本为目标平台的应用程序，在默认情况下无法访问属于其他应用的专属目录。

由于外部存储可能会随时被用户从设备上移除，因此，在尝试从外部存储读取数据或将数据写入外部存储之前，需要首先验证其是否能够正常访问。可以通过 Environment.getExternalStorageState()方法查询外置存储卡的状态，如果返回的状态为 MEDIA_MOUNTED，则说明存储卡状态正常，可以在外部存储中正常读取和写入数据文件。如果返回的状态为 MEDIA_MOUNTED_READ_ONLY，则说明外部存储当前是只读状态，只能执行读操作而无法写入数据。如果返回的是其他状态，则说明当前外部存储状态不正常，不能对其执行正常的读/写操作。

例如，以下方法可用于确定存储空间的可用性：

```
    private boolean isExternalStorageWritable() {
        return Environment.getExternalStorageState() ==
    Environment.MEDIA_MOUNTED;
    }

    private boolean isExternalStorageReadable() {
        return Environment.getExternalStorageState() ==
    Environment.MEDIA_MOUNTED ||
        Environment.getExternalStorageState() ==
    Environment.MEDIA_MOUNTED_READ_ONLY;
    }
```

（1）访问持久性文件。

如果从外部存储空间访问应用专属文件，则需使用 getExternalFilesDir()方法，其代码如下。

```
    File appSpecificExternalFile = new File(context.getExternalFilesDir(), filename);
```

获取到代表外部存储文件的 File 对象之后，可以进一步通过文件读/写的 Java I/O 类执行相关读/写操作。

（2）访问缓存文件。

如果将文件添加到外部存储空间中的缓存，则需获取对 externalCacheDir 的引用：
```
File externalCacheFile = new File(context.getExternalCacheDir(), filename);
```
如果从外部缓存目录中移除文件，则直接使用该文件的 File 对象的 delete()方法：
```
externalCacheFile.delete();
```

2．外部存储中共享存储空间的访问

如果当前应用程序要存储的是公共数据，允许被别的应用程序访问，或者希望在应用卸载后数据仍然保留，这时就需要把数据存储在外部存储器的公共空间。

Android 提供了用于存储和访问以下类型可共享数据的相关类库。

（1）媒体内容。系统提供标准的公共目录来存储这些类型的文件，每种媒体文件都有自己专属的存储目录。即使应用已卸载，这些文件仍会保留在用户的设备上。应用程序可以使用 MediaStore 相关的 API 访问这些内容。

系统会自动扫描外部存储卷，并将媒体文件添加到以下明确定义的集合中。

① 图片（包括照片和屏幕截图）存储在 DCIM/和 Pictures/目录中。系统将这些文件添加到 MediaStore.Images 表格。

② 视频存储在 DCIM/、Movies/和 Pictures/目录中。系统将这些文件添加到 MediaStore.Video 表格。

③ 音频文件存储在 Alarms/、Audiobooks/、Music/、Notifications/、Podcasts/和 Ringtones/目录中，以及位于 Music/或 Movies/目录的音频播放列表中。系统将这些文件添加到 MediaStore.Audio 表格。

④ 下载文件存储在 Downloads/目录中。在搭载 Android 10 及更高版本系统的设备上，这些文件存储在 MediaStore.Downloads 目录中。此目录在 Android 9 及更低版本的系统中不可用。

以 Android 10 或更高版本为目标平台的应用中，媒体库还包含一个名为 MediaStore.Files 的集合。

（2）文档和其他文件。系统有一个特殊目录，用于包含其他文件类型，如 PDF、EPUB 等格式的文件。应用程序可以使用此平台的存储访问框架（Storage Access Framework）访问这些文件。

访问外部存储中的共享文件，需要经过以下两个步骤。

① 请求必要权限。

在对媒体文件执行操作之前，需要确保当前程序已声明访问这些文件所需的权限。

用于访问应用内媒体文件的权限模型取决于应用程序是否使用分区存储（仅适用于以 Android 10 或更高版本为目标平台的应用程序）。如果应用使用分区存储，则应仅针对搭载 Android 9（API 级别 28）或更低版本的设备请求存储相关权限。通过将 android:maxSdkVersion 属性添加到应用清单文件中的权限声明来应用此条件：
```
<uses-permission
    android:name="android.permission.WRITE_EXTERNAL_STORAGE"
    android:maxSdkVersion="28" />
```

在 Android 10 或更高版本的系统上，应用程序访问自己创建的文件是不需要声明任何文件权限的。只有当读/写其他应用在外部共享文件夹中创建的文件时，才需要根据所做的操作

在清单文件中声明读或写的权限,并在程序执行相关操作之前向用户动态申请所需权限。

请勿为搭载 Android 10 或更高版本的设备请求不必要的存储相关权限。应用可以提供明确定义的媒体集合,包括 MediaStore.Downloads 集合。例如,如果用户正在开发一款相机应用,则无须请求存储相关权限,因为应用中已写入媒体库的图片。

② 文件访问。

对外部共享文件夹的访问,需要使用 MediaStore 相关类库来访问,可以完成针对各类媒体文件和文档的添加、删除、内容修改等常见操作。

5.3.4 文件存储操作示例

【例 5.3】开发一款应用程序,实现文件内部存储、外部私有数据存储和外部公有数据存储的功能。

通过三个界面分别演示三种文件存储方式,主界面如图 5.6 所示。这三种存储方式的操作界面分别如图 5.7、图 5.8 和图 5.9 所示。在内部存储和外部私有存储界面中,单击"写文件"按钮,将用户输入的文字分别通过内部和外部私有存储置于内部存储区域或 SD 卡的数据文件内;单击"读文件"按钮,将读取上一步写入的内容,显示在 TextView 控件中。在外部公共存储界面,放置了一个 ImageView 控件,单击"创建"按钮将会使用 MediaStore 在外部公共存储中创建一个图片文件,并显示在 ImageView 中;单击"查询"按钮,程序会从外部公共存储中读取已创建的图片文件,并显示在 ImageView 控件中,如图 5.10 所示。

图 5.6 文件存储主界面 　　 图 5.7 内部存储界面 　　 图 5.8 外部私有存储

文件的内部存储和外部私有存储操作均不需要权限,但外部公共存储操作则需要 WRITE_EXTERNAL_STORAGE 和 READ_EXTERNAL_STORAGE 权限。

(1) 权限的声明。

创建一个应用程序,在清单文件 AndroidManifest.xml 中,声明如下权限,用于兼容 Android 9 及更低版本系统操作文件外部存储。

```
<uses-permission
    android:name="android.permission.WRITE_EXTERNAL_STORAGE"
```

```
            android:maxSdkVersion="28" />
        <uses-permission
            android:name="android.permission.READ_EXTERNAL_STORAGE"
            android:maxSdkVersion="28" />
```

 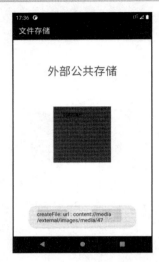

图 5.9 文件外部公共存储初始界面　　图 5.10 文件外部公共存储操作结果

（2）创建主界面。

在 MainActivity 对应的布局文件 activity_main.xml 中添加一个 TextView 控件和三个 Button 控件，代码如下：

```xml
<?xml version="1.0" encoding="utf-8"?>
<androidx.constraintlayout.widget.ConstraintLayout
    xmlns:android="http://schemas.android.com/apk/res/android"
    xmlns:app="http://schemas.android.com/apk/res-auto"
    xmlns:tools="http://schemas.android.com/tools"
    android:layout_width="match_parent"
    android:layout_height="match_parent"
    tools:context=".MainActivity">
    <TextView
        android:id="@+id/tvTitle"
        android:layout_width="251dp"
        android:layout_height="66dp"
        android:layout_marginTop="44dp"
        android:gravity="center"
        android:lineSpacingExtra="24sp"
        android:text="文件存储"
        android:textColor="@color/purple_500"
        android:textSize="40sp"
        android:textStyle="bold"
        app:layout_constraintLeft_toLeftOf="parent"
        app:layout_constraintRight_toRightOf="parent"
        app:layout_constraintTop_toTopOf="parent" />
    <Button
        android:id="@+id/btnInternal"
```

```xml
            android:layout_width="wrap_content"
            android:layout_height="wrap_content"
            android:layout_marginTop="92dp"
            android:textSize="30dp"
            android:text="内部存储"
            app:layout_constraintEnd_toEndOf="parent"
            app:layout_constraintStart_toStartOf="parent"
            app:layout_constraintTop_toBottomOf="@+id/tvTitle" />
    <Button
            android:id="@+id/btnExternalPrivate"
            android:layout_width="wrap_content"
            android:layout_height="wrap_content"
            android:layout_marginTop="56dp"
            android:textSize="30dp"
            android:text="外部私有存储"
            app:layout_constraintEnd_toEndOf="parent"
            app:layout_constraintHorizontal_bias="0.498"
            app:layout_constraintStart_toStartOf="parent"
            app:layout_constraintTop_toBottomOf="@+id/btnInternal" />
    <Button
            android:id="@+id/btnExternalPublic"
            android:layout_width="wrap_content"
            android:layout_height="wrap_content"
            android:layout_marginTop="40dp"
            android:textSize="30dp"
            android:text="外部公共存储"
            app:layout_constraintEnd_toEndOf="@+id/btnExternalPrivate"
            app:layout_constraintHorizontal_bias="0.428"
            app:layout_constraintStart_toStartOf="@+id/btnExternalPrivate"
            app:layout_constraintTop_toBottomOf="@+id/btnExternalPrivate" />
</androidx.constraintlayout.widget.ConstraintLayout>
```

（3）创建内部存储界面。

在应用程序中创建一个 Activity，命名为 InternalStorage，对应布局文件 activity_internal_storage.xml 的代码如下：

```xml
<?xml version="1.0" encoding="utf-8"?>
<androidx.constraintlayout.widget.ConstraintLayout
    xmlns:android="http://schemas.android.com/apk/res/android"
    xmlns:app="http://schemas.android.com/apk/res-auto"
    android:layout_width="match_parent"
    android:layout_height="match_parent"
    xmlns:tools="http://schemas.android.com/tools"
    tools:context=".InternalStorage">
    <TextView
        android:id="@+id/tvTitle"
        android:layout_width="409dp"
        android:layout_height="wrap_content"
        android:layout_marginTop="104dp"
        android:text="内部存储"
        android:gravity="center"
        android:textSize="30sp"
        app:layout_constraintEnd_toEndOf="parent"
```

```xml
        app:layout_constraintStart_toStartOf="parent"
        app:layout_constraintTop_toTopOf="parent" />
    <EditText
        android:id="@+id/etContent"
        android:layout_width="wrap_content"
        android:layout_height="wrap_content"
        android:layout_marginLeft="28dp"
        android:layout_marginTop="40dp"
        android:ems="10"
        android:hint="input message"
        android:inputType="text"
        app:layout_constraintStart_toStartOf="parent"
        app:layout_constraintTop_toBottomOf="@+id/tvTitle" />
    <Button
        android:id="@+id/btnWrite"
        android:layout_width="wrap_content"
        android:layout_height="wrap_content"
        android:layout_marginTop="40dp"
        android:layout_marginRight="44dp"
        android:textSize="20sp"
        android:text="写文件"
        app:layout_constraintEnd_toEndOf="parent"
        app:layout_constraintTop_toBottomOf="@+id/tvTitle" />
    <Button
        android:id="@+id/btnRead"
        android:layout_width="wrap_content"
        android:layout_height="wrap_content"
        android:layout_marginTop="36dp"
        android:textSize="20sp"
        android:text="读文件"
        app:layout_constraintStart_toStartOf="@+id/btnWrite"
        app:layout_constraintTop_toBottomOf="@+id/btnWrite" />
    <TextView
        android:id="@+id/tvContent"
        android:layout_width="210dp"
        android:layout_height="46dp"
        android:layout_marginLeft="28dp"
        android:layout_marginTop="40dp"
        android:textSize="20sp"
        android:background="#EAE1E1"
        android:gravity="start|center_vertical"
        app:layout_constraintStart_toStartOf="parent"
        app:layout_constraintTop_toBottomOf="@+id/etContent"/>
</androidx.constraintlayout.widget.ConstraintLayout>
```

（4）创建外部私有存储界面。

在应用程序中创建一个 Activity，命名为 ExternalPrivateStorage，对应布局文件 activity_external_private_storage.xml 的代码如下：

```xml
<?xml version="1.0" encoding="utf-8"?>
<androidx.constraintlayout.widget.ConstraintLayout
    xmlns:android="http://schemas.android.com/apk/res/android"
    xmlns:app="http://schemas.android.com/apk/res-auto"
```

```xml
xmlns:tools="http://schemas.android.com/tools"
android:layout_width="match_parent"
android:layout_height="match_parent"
tools:context=".ExternalPrivateStorage">
<TextView
    android:id="@+id/tvTitle"
    android:layout_width="409dp"
    android:layout_height="wrap_content"
    android:layout_marginTop="104dp"
    android:gravity="center"
    android:text="外部私有存储"
    android:textSize="30sp"
    app:layout_constraintEnd_toEndOf="parent"
    app:layout_constraintStart_toStartOf="parent"
    app:layout_constraintTop_toTopOf="parent" />
<EditText
    android:id="@+id/etContent"
    android:layout_width="wrap_content"
    android:layout_height="wrap_content"
    android:layout_marginLeft="28dp"
    android:layout_marginTop="40dp"
    android:ems="10"
    android:hint="input message"
    android:inputType="text"
    app:layout_constraintStart_toStartOf="parent"
    app:layout_constraintTop_toBottomOf="@+id/tvTitle" />
<Button
    android:id="@+id/btnWrite"
    android:layout_width="wrap_content"
    android:layout_height="wrap_content"
    android:layout_marginTop="36dp"
    android:layout_marginEnd="36dp"
    android:layout_marginRight="36dp"
    android:text="写文件"
    android:textSize="20sp"
    app:layout_constraintEnd_toEndOf="parent"
    app:layout_constraintTop_toBottomOf="@+id/tvTitle" />
<Button
    android:id="@+id/btnRead"
    android:layout_width="wrap_content"
    android:layout_height="wrap_content"
    android:layout_marginTop="36dp"
    android:text="读文件"
    android:textSize="20sp"
    app:layout_constraintStart_toStartOf="@+id/btnWrite"
    app:layout_constraintTop_toBottomOf="@+id/btnWrite" />
<TextView
    android:id="@+id/tvContent"
    android:layout_width="210dp"
    android:layout_height="46dp"
    android:layout_marginTop="40dp"
    android:gravity="start|center_vertical"
    android:text="TextView"
```

```
                android:textSize="20sp"
                app:layout_constraintEnd_toEndOf="@+id/etContent"
                app:layout_constraintHorizontal_bias="0.0"
                app:layout_constraintStart_toStartOf="@+id/etContent"
                app:layout_constraintTop_toBottomOf="@+id/etContent" />
        </androidx.constraintlayout.widget.ConstraintLayout>
```

（5）创建外部公共存储界面。

在应用程序中创建一个 Activity，命名为 External_Storage_Public，对应布局文件 activity_external_public_storage.xml 的代码如下：

```
        <?xml version="1.0" encoding="utf-8"?>
        <androidx.constraintlayout.widget.ConstraintLayout
            xmlns:android="http://schemas.android.com/apk/res/android"
            xmlns:app="http://schemas.android.com/apk/res-auto"
            xmlns:tools="http://schemas.android.com/tools"
            android:layout_width="match_parent"
            android:layout_height="match_parent"
            tools:context=".ExternalPublicStorage">
            <ImageView
                android:id="@+id/iv1"
                android:layout_width="200dp"
                android:layout_height="200dp"
                android:scaleType="centerInside"
                app:layout_constraintBottom_toBottomOf="parent"
                app:layout_constraintEnd_toEndOf="parent"
                app:layout_constraintHorizontal_bias="0.497"
                app:layout_constraintStart_toStartOf="parent"
                app:layout_constraintTop_toTopOf="parent"
                app:layout_constraintVertical_bias="0.472" />
            <TextView
                android:id="@+id/tvTitle"
                android:layout_width="335dp"
                android:layout_height="48dp"
                android:layout_marginTop="72dp"
                android:text="外部公共存储"
                android:gravity="center"
                android:textSize="30sp"
                app:layout_constraintEnd_toEndOf="parent"
                app:layout_constraintStart_toStartOf="parent"
                app:layout_constraintTop_toTopOf="parent" />
            <Button
                android:id="@+id/btnCreate"
                android:layout_width="wrap_content"
                android:layout_height="wrap_content"
                android:layout_marginLeft="72dp"
                android:layout_marginTop="124dp"
                android:textSize="20sp"
                android:text="创建"
                app:layout_constraintStart_toStartOf="parent"
                app:layout_constraintTop_toBottomOf="@+id/iv1" />
            <Button
                android:id="@+id/btnQuery"
```

```
            android:layout_width="wrap_content"
            android:layout_height="wrap_content"
            android:layout_marginLeft="60dp"
            android:layout_marginTop="124dp"
            android:textSize="20sp"
            android:text="查询"
            app:layout_constraintEnd_toEndOf="parent"
            app:layout_constraintHorizontal_bias="0.129"
            app:layout_constraintStart_toEndOf="@+id/btnCreate"
            app:layout_constraintTop_toBottomOf="@+id/iv1" />
</androidx.constraintlayout.widget.ConstraintLayout>
```

（6）编写 MainActivity。

在主界面上单击三个按钮可以分别跳转到不同存储方式的操作界面。MainActivity.java 具体代码如下：

```java
public class MainActivity extends AppCompatActivity implements View.OnClickListener {
    Button btnInternal,btnExternalPrivate,btnExternalPublic;
    @Override
    protected void onCreate(Bundle savedInstanceState) {
        super.onCreate(savedInstanceState);
        setContentView(R.layout.activity_main);
        btnInternal= findViewById(R.id.btnInternal);
        btnExternalPrivate = findViewById(R.id.btnExternalPrivate);
        btnExternalPublic = findViewById(R.id.btnExternalPublic);
        btnInternal.setOnClickListener(this);
        btnExternalPrivate.setOnClickListener(this);
        btnExternalPublic.setOnClickListener(this);
    }
    @Override
    public void onClick(View v) {
        Intent intent = new Intent();
        switch (v.getId()) {
            case R.id.btnInternal:
                intent.setClass(this, InternalStorage.class);
                break;
            case R.id.btnExternalPrivate:
                intent.setClass(this, ExternalPrivateStorage.class);
                break;
            case R.id.btnExternalPublic:
                intent.setClass(this, ExternalPublicStorage.class);
                break;
        }
        startActivity(intent);
    }
}
```

（7）内部存储实现。

单击主界面上的"内部存储"按钮，跳转到内部存储界面。在该界面的编辑框内输入任意字符串，单击"写文件"按钮，将输入的内容写入内部存储；单击"读文件"按钮，从内部存储读取文件内容，并显示于界面上。具体代码如下：

```java
public class InternalStorage extends AppCompatActivity implements View.OnClickListener{
    Button btnRead, btnWrite;
    EditText etContent;
    TextView tvContent;
    @Override
    protected void onCreate(Bundle savedInstanceState) {
        super.onCreate(savedInstanceState);
        setContentView(R.layout.activity_internal_storage);
        etContent = findViewById(R.id.etContent);
        btnRead = findViewById(R.id.btnRead);
        btnWrite = findViewById(R.id.btnWrite);
        tvContent = findViewById(R.id.tvContent);
        btnRead.setOnClickListener(this);
        btnWrite.setOnClickListener(this);
    }
    @Override
    public void onClick(View v) {
        switch (v.getId()) {
            case R.id.btnRead:
                String content = readData();
                if (content != null){
                    tvContent.setText(content);
                }
                break;
            case R.id.btnWrite:
                String msg = etContent.getText().toString();
                saveData(msg);
                break;
        }
    }
    public void saveData(String data) {   //写文件
        try {
            FileOutputStream fos = openFileOutput("test.txt", MODE_PRIVATE);
            fos.write(data.getBytes());
            fos.close();
        }
        catch (IOException ex)
        {
            ex.printStackTrace();
        }
    }
    public String readData() {   //读文件
        try {
            FileInputStream fis = openFileInput("test.txt");
            BufferedReader br =   new BufferedReader( new InputStreamReader(fis));
            String content = br.readLine();
            return content;
        }
        catch (IOException ex){
            ex.printStackTrace();
            return null;
```

            }
        }
    }

(8) 外部私有存储实现。

单击主界面上的"外部私有存储"按钮，跳转到外部私有存储界面。在该界面的编辑框内输入任意字符串，单击"写文件"按钮，将输入的内容写入外部私有存储，单击"读文件"按钮，从外部私有存储读出文件内容，并显示于界面上。具体代码如下：

```java
public class ExternalPrivateStorage extends AppCompatActivity implements View.OnClickListener {
    Button btnRead, btnWrite;
    EditText etContent;
    TextView tvContent;
    @Override
    protected void onCreate(Bundle savedInstanceState) {
        super.onCreate(savedInstanceState);
        setContentView(R.layout.activity_external_private_storage);
        etContent = findViewById(R.id.etContent);
        btnRead = findViewById(R.id.btnRead);
        btnWrite = findViewById(R.id.btnWrite);
        tvContent = findViewById(R.id.tvContent);
        btnRead.setOnClickListener(this);
        btnWrite.setOnClickListener(this);
    }
    @Override
    public void onClick(View v) {
        switch (v.getId()) {
            case R.id.btnRead:
                String content = readData();
                if (content != null) {
                    tvContent.setText(content);
                }
                break;
            case R.id.btnWrite:
                String msg = etContent.getText().toString();
                saveData(msg);
                break;
        }
    }
    // saveData()方法将内容存储到外部私有文件中
    public void saveData(String data) {
        try {
            String state = Environment.getExternalStorageState();
            if (state.equals(Environment.MEDIA_MOUNTED)) {
                File sd = getExternalFilesDir(null);
                File dataFile = new File(sd, "data.txt");
                FileOutputStream fos = new FileOutputStream(dataFile);
                fos.write(data.getBytes());
                fos.close();
            } else {
                Toast.makeText(this, "write error!", Toast.LENGTH_LONG).show();
            }
        } catch (IOException ex) {
```

```java
            ex.printStackTrace();
        }
    }
    // readData()方法从外部私有存储读文件
    public String readData() {
        try {
            String state = Environment.getExternalStorageState();
            if (state.equals(Environment.MEDIA_MOUNTED) || state.equals(Environment.MEDIA_MOUNTED_READ_ONLY)) {
                File sd = getExternalFilesDir(null);
                File dataFile = new File(sd, "data.txt");
                FileInputStream fis = new FileInputStream(dataFile);
                BufferedReader br = new BufferedReader(new InputStreamReader(fis));
                String msg = br.readLine();
                return msg;
            } else {
                Toast.makeText(this, "read error!", Toast.LENGTH_LONG).show();
                return null;
            }
        } catch (IOException ex) {
            ex.printStackTrace();
            return null;
        }
    }
}
```

（9）外部公共存储实现。

在主界面上单击"外部公共存储"按钮，跳转到外部公共存储界面。单击该界面上的"创建"按钮，创建一个图片文件，使用 MediaStore 存储于外部公共存储之中，并在界面上显示该图片。单击"查询"按钮，可从外部公共存储中查找上一步创建的图片，并显示于界面上。具体代码如下：

```java
public class ExternalPublicStorage extends AppCompatActivity implements View.OnClickListener {
    final int REQUEST_PERMISSION = 100;
    final String TAG = this.getClass().toString();
    ImageView imageView ;
    Button btnQuery, btnCreate;
    @Override
    protected void onCreate(Bundle savedInstanceState) {
        super.onCreate(savedInstanceState);
        setContentView(R.layout.activity_external_public_storage);
        imageView = findViewById(R.id.iv1);
        btnCreate = findViewById(R.id.btnCreate);
        btnQuery = findViewById(R.id.btnQuery);
        btnCreate.setOnClickListener(this);
        btnQuery.setOnClickListener(this);
    }
    @Override
    public void onRequestPermissionsResult(int requestCode, @NonNull String[] permissions, @NonNull int[] grantResults) {
        super.onRequestPermissionsResult(requestCode, permissions, grantResults);
        if (requestCode == REQUEST_PERMISSION && grantResults[0] == PackageManager.PERMISSION_GRANTED) {
```

```java
                    Toast.makeText(this, "已获得 SD 卡写权限", Toast.LENGTH_LONG).show();
                }
            }
            private void grantPermission() {
                // 使用 Environment.isExternalStorageLegacy()来检查 App 的运行模式
                if (Build.VERSION.SDK_INT >= Build.VERSION_CODES.Q && !Environment.isExternalStorageLegacy()) {
                    System.out.println("btnCreate = isExternalStorageLegacy : " + Environment.isExternalStorageLegacy());
                }
                if (ActivityCompat.checkSelfPermission(this, Manifest.permission.WRITE_EXTERNAL_STORAGE) != PackageManager.PERMISSION_GRANTED) {
                    ActivityCompat.requestPermissions(this, new String[]{Manifest.permission.WRITE_EXTERNAL_STORAGE},
                            REQUEST_PERMISSION);
                    Log.d(TAG, "grantPermission: 权限说明提示");
                } else {
                    if (Build.VERSION.SDK_INT >= Build.VERSION_CODES.M) {
                        Log.d(TAG, "grantPermission: requestPermissions");
                        requestPermissions(new String[]{Manifest.permission.WRITE_EXTERNAL_STORAGE}, REQUEST_PERMISSION);
                    }
                }
            }
            private void createFile() {   //创建文件
                Uri contentUri = MediaStore.Images.Media.getContentUri(MediaStore.VOLUME_EXTERNAL);
                ContentValues contentValues = new ContentValues();
                long dateTaken = System.currentTimeMillis();
                contentValues.put(MediaStore.Images.Media.DATE_TAKEN, dateTaken);
                contentValues.put(MediaStore.Images.Media.DESCRIPTION, "创建的第一张图片");
                contentValues.put(MediaStore.Images.Media.IS_PRIVATE, 1);
                contentValues.put(MediaStore.Images.Media.DISPLAY_NAME, "test.png");
                contentValues.put(MediaStore.Images.Media.MIME_TYPE, "image/png");
                contentValues.put(MediaStore.Images.Media.TITLE, "图片");
                contentValues.put(MediaStore.Images.Media.RELATIVE_PATH, "DCIM/test");
                long dateAdded = System.currentTimeMillis();
                contentValues.put(MediaStore.Images.Media.DATE_ADDED, dateAdded);
                long dateModified = System.currentTimeMillis();
                contentValues.put(MediaStore.Images.Media.DATE_MODIFIED, dateModified);

                Uri insertUri = getContentResolver().insert(contentUri, contentValues);
                Log.d(TAG, "createFile: url : " + insertUri);
                Toast.makeText(this, "createFile: url : " + insertUri, Toast.LENGTH_LONG).show();
                try {
                    OutputStream outputStream = getContentResolver().openOutputStream(insertUri);
                    Bitmap bitmap = Bitmap.createBitmap(600, 600, Bitmap.Config.ARGB_8888);
                    Canvas canvas = new Canvas(bitmap);
                    canvas.drawColor(Color.RED);
                    Paint paint = new Paint();
                    paint.setColor(Color.BLACK);
                    paint.setTextSize(40);
                    canvas.drawText("创建的图片", 100, 100, paint);
```

```java
                    bitmap.compress(Bitmap.CompressFormat.PNG, 90, outputStream);
                    outputStream.close();
                } catch (Exception e) {
                    e.printStackTrace();
                } finally {
                    showImg(insertUri);
                }
            }
        }

        private Uri selectSingle() {
            Uri queryUri = null;
            Uri external = MediaStore.Images.Media.EXTERNAL_CONTENT_URI;
            String selection = MediaStore.Images.Media.DISPLAY_NAME + "=?";
            String[] args = new String[]{"test.png"};
            String[] projection = new String[]{MediaStore.Images.Media._ID};
            Cursor cursor = getContentResolver().query(external, projection, selection, args, null);
            if (cursor != null && cursor.moveToFirst()) {
                queryUri = ContentUris.withAppendedId(external, cursor.getLong(0));
                Log.d(TAG, "selectSingle 查询成功, Uri 路径：" + queryUri);
                Toast.makeText(this, queryUri.toString(), Toast.LENGTH_LONG).show();
                showImg(queryUri);
                cursor.close();
                return queryUri;
            } else {
                Log.d(TAG, "selectSingle 查询失败");
                return null;
            }
        }

        private void deleteFile() {
            System.out.println("MediaStore.Images.Media.DISPLAY_NAME = " + MediaStore.Images.Media.DISPLAY_NAME);

            Uri CONTENT_URI = MediaStore.Images.Media.EXTERNAL_CONTENT_URI;
            String selectionclause = MediaStore.Images.Media.DISPLAY_NAME + "=?";
            String[] arguments = new String[]{"test.png"};
            int delete = getContentResolver().delete(CONTENT_URI, selectionclause, arguments);
            Log.d(TAG, "deleteFile: " + delete);
            if (delete > -1) {
                //如果发生异常返回-1
                Toast.makeText(this, "delete : " + delete, Toast.LENGTH_LONG).show();
            } else {
                Log.d(TAG, "deleteFile: 失败 " + delete);
            }
        }
```

5.4 数据库 SQLite

当 Android 程序中需要存储较多的结构化数据时，一般考虑将其优先存入数据库中。SQLite 是 Android 系统中内置的轻量级数据库系统，本节将重点介绍如何使用程序实现 SQLite 数据库的创建、数据表的创建，以及对数据表的增、删、改、查等操作。

5.4.1 SQLite 数据库简介

SQLite 是一款轻型的嵌入式数据库，由于其占用资源低（占用内存只需几百 KB）、处理速度快等特点，目前许多嵌入式产品都使用了它。SQLite 能够支持 Windows/Linux/UNIX 等主流的操作系统，同时能够与很多程序语言相结合，如 Tcl、C#、PHP、Java 等，还有 ODBC 接口，同样比 MySQL、PostgreSQL 这两款开源数据库管理系统的处理速度都快。SQLite 数据库具有如下特点。

（1）轻量级。

使用 SQLite 只需要一个动态库就可以享受其全部功能。

（2）独立性。

SQLite 数据库的核心引擎不需要依赖第三方软件，也不需要进行所谓的"安装"操作。

（3）隔离性。

SQLite 数据库中所有的信息（如表、视图、触发器等）都包含在一个文件夹内，以方便管理和维护。

（4）跨平台。

SQLite 目前支持大部分操作系统，不仅是计算机操作系统，而且在众多的手机系统和各种嵌入式设备上也可以运行，如 Android 和 iOS。

（5）多语言接口。

SQLite 数据库支持多语言编程接口。

（6）安全性。

SQLite 数据库通过数据库级上的独占性和共享锁来实现独立事务处理。这意味着多个进程可以同时从同一个数据库中读取数据，但只能有一个可以写入数据。

5.4.2 创建 SQLite 数据库

使用 SQLite 数据库存储数据，首先需要在 Android 应用程序中创建数据库。在 Android 系统中，一个应用程序创建的数据库通常保存在/data/data/应用包名/database 目录下，所有数据均保存在一个数据库文件中。

创建数据库的工作通常通过 Android SDK 提供的 SQLiteOpenHelper 类来完成，这是一个抽象类，使用时需要创建其子类，并且至少需要实现以下三种方法。

（1）构造方法。

public SQLiteOpenHelper(Context context,String name,CursorFactory factory,int version);

其中，name 参数表示数据库文件名（不包括文件路径），SQLiteOpenHelper 类会根据这个文件名创建数据库文件。version 表示数据库的版本号。如果当前传入的数据库版本号比上次创建或升级的版本号高，SQLiteOpenHelper 类就会调用 onUpgrage ()方法更新数据库的版本号。数据库第一次创建时会有一个初始的版本号，当需要对数据库的表、视图等升级时可以先增大版本号，再进行重新创建。

（2）onCreate()方法。

public abstract void onCreate(SQLiteDatabase db);

该方法在数据库文件第一次创建时调用，在此方法中可进行创建表结构、视图等操作。

（3）onUpgrage()方法。

```
public abstract void onUpdate(SQLiteDatabase db,int oldVersion,int newVersion);
```

在数据库版本升高时调用 onUpgrage()方法，可以在该方法中根据新旧版本号进行删除表、添加表、修改表等操作。

如果数据库文件不存在，SQLiteOpenHelper 在自动创建数据库后会调用 onCreate()方法。如果数据库文件存在，并且当前版本号高于上次创建或升级的版本号，SQLiteOpenHelper 就会调用 onUpgrage ()方法，调用该方法后会更新数据库的版本号。

使用下面的代码创建一个 SQLiteOpenHelper 的子类：

```java
public class MySQLiteHelper extends SQLiteOpenHelper {
    //数据库名称和表名称
    private static final String DB_NAME="student.db";
    private static final String TABLE_NAME="studentScore";
    //数据库版本号
    private static int VERSION = 1;
    //表中各字段名称
    private static final String STU_ID="_id";
    private static final String STU_NAME="name";
    private static final String STU_SCORE="score";
    //创建表的 SQL 语句
    private static final String CREATE_TABLE="create table   "+TABLE_NAME+
        " ( "+ STU_ID+"   integer primary key autoincrement,"
        + STU_NAME+"   text not null,"
        +STU_SCORE+"   integer );";
    public MySQLiteHelper (Context context, String name, SQLiteDatabase CursorFactory factory, int version) {
        //必须通过 super 调用父类的构造函数
        super(context, name, factory, version);
    }
    public MySQLiteHelper (Context context){
        super(context,DB_NAME,null, VERSION);
    }
    @Override
    public void onCreate(SQLiteDatabase db) {
        db.execSQL(CREATE_TABLE);   //执行 SQL 语句创建表
    }
    @Override
    public void onUpgrade(SQLiteDatabase db, int oldVersion, int newVersion) {
        db.execSQL("DROP TABLE IF EXISTS "+ TABLE_NAME);
        onCreate(db);
    }
}
```

5.4.3 数据库操作的实现

创建 SQLiteOpenHelper 的子类后，通过调用该类的构造方法可生成一个 SQLiteOpenHelper

对象，然后通过 SQLiteOpenHelper 对象的 getReadableDatabase()方法或 getWriteableDatabase()方法可以获取一个 SQLiteDatabase 对象，使用 SQLiteDatabase 对象提供的 insert()、query()、update()、delete()等方法，可以实现对数据库的增、删、改、查等常用操作。

1．SQLiteDatabase 类的常用方法

（1）insert()方法。

```
public long insert (String table, String   nullColumnHack, ContentValues values)
```

该方法向指定的数据表中增加一行数据，并返回一个 Long 型的值，表示被增加行的 ID。如果返回为–1，则表示增加失败。方法中各参数的含义如下：

- table：要增加数据的表名称。
- nullColumnHack：空值字段的名称。
- Values：ContentValues 对象，指包含被增加表中一行数据的各字段值。

（2）query()方法。

```
public Cursor query (boolean distinct, String table, String[] columns,
    String selection, String[] selectionArgs, String groupBy,
    String having, String orderBy, String limit)
```

该方法可查询指定的数据表，并返回一个带游标的数据集，其参数含义如下。

- distinct：布尔类型的可选项，用来说明返回值是否只包含唯一的值。
- table：要查询数据的表名称。
- columns：由列名构成的数组。
- selection：条件 where 子句，可包含"？"通配符，在子句中用于占位符。
- selectionArgs：参数数组，可替换 where 子句中的"？"占位符。
- groupBy：分组列。
- having：分组条件。
- orderBy：排序列。
- limit：可选项，用来对返回的行数进行限制。

（3）update()方法。

```
public int update(String table, ContentValues values, String whereClause, String[] whereArgs)
```

该方法可修改指定数据表中的数据，并返回一个 Int 类的值，用于表示被修改的行数。各参数的含义如下。

- table：需要修改数据的表名称。
- values：包含要修改的字段和值。
- whereClause：修改条件。
- whereArgs：修改条件 whereClause 需要的参数数组。

（4）delete()方法。

```
public int delete(String table, String whereClause, String[] whereArgs)
```

该方法可删除指定数据表中的数据，并返回一个 Int 类型的值，用于表示被删除的行数。各参数的含义如下。

- table：需要删除数据的表名称。
- whereClause：删除条件。

- whereArgs：删除条件所需的参数数组。

(5) execSQL()方法。

```
public void execSQL (String sql)
```

该方法用于执行除查询和其他返回数据的 SQL 语句之外的所有 SQL 语句。

(6) rawQuery()方法。

```
public Cursor rawQuery (String sql, String[] selectionArgs, CancellationSignal cancellationSignal)
```

该方法用于执行带占位符的 SQL 查询语句，参数 selectionArgs 用来逐一替代 SQL 中的占位符。

(7) close()方法。

```
public void close()
```

该方法用于关闭数据库的连接对象。

2．查询结果的处理

无论是通过 query()方法还是使用 rawQuery()方法执行查询操作，最终都会返回一个包含查询结果的 Cursor 对象。Cursor 的常用方法如表 5.4 所示。

表 5.4 Cursor 的常用方法

方　　法	功　能　描　述
public abstract boolean moveToFirst ()	移动游标到第一行
public abstract boolean moveToLast()	移动游标到最后一行
moveToNext()	移动游标到下一行
moveToPrevious()	移动游标到前一行
isFirst()	判断当前行是否是第一行
isLast()	判断当前行是否是最后一行
isBeforeFirst()	游标是否在第一行之前
isAfterLast()	游标是否在最后一行之后
getCount()	获取查询结果的总条数
getColumnCount()	获取列数
getColumnIndex(String columnName)	由字段名获取列的索引
getColumnNames()	获取所有列名
getColumnName(int columnIndex)	由字段索引获取列名

Cursor 就像一个迭代器，要获取每一行数据就要将其放置在一个 while 循环中，在每次循环中光标将下移一行。随后在 while 循环当中，将需要读取数据的列索引传递到 Cursor 的 getXXX()方法中获取该列的字段值。

5.4.4　SQLite 数据库使用示例

【例 5.4】开发一个学生成绩管理程序，用 SQLite 数据库存储学生成绩，并实现对学生成绩的增、删、改、查操作，操作界面如图 5.11 所示。按学号进行修改和删除，将操作结果显示在界面下方的编辑框内。

图 5.11　操作界面

（1）界面设计。

创建一个应用程序，编辑布局文件，布局代码如下：

```xml
<?xml version="1.0" encoding="utf-8"?>
<LinearLayout xmlns:android="http://schemas.android.com/apk/res/android"
    android:orientation="vertical"
    android:padding="20sp"
    android:layout_width="match_parent"
    android:layout_height="match_parent">
    <LinearLayout
        android:layout_width="match_parent"
        android:layout_height="wrap_content"
        android:orientation="horizontal">
        <TextView
            android:id="@+id/textView"
            android:layout_width="0dp"
            android:layout_height="wrap_content"
            android:layout_weight="1"
            android:gravity="center"
            android:textSize="30sp"
            android:text="学号" />
        <EditText
            android:id="@+id/edit_num"
            android:layout_width="0dp"
            android:layout_height="wrap_content"
            android:layout_weight="3"
            android:textSize="30sp"
            android:inputType="textPersonName"
            android:text="" />
    </LinearLayout>
    <LinearLayout
        android:layout_width="match_parent"
        android:layout_height="wrap_content"
        android:orientation="horizontal">
```

```xml
<TextView
    android:id="@+id/textView2"
    android:layout_width="0dp"
    android:layout_height="wrap_content"
    android:layout_weight="1"
    android:gravity="center"
    android:textSize="30sp"
    android:text="姓名" />
<EditText
    android:id="@+id/edit_name"
    android:layout_width="0dp"
    android:layout_height="wrap_content"
    android:layout_weight="3"
    android:textSize="30sp"
    android:inputType="textPersonName"
    android:text="" />
</LinearLayout>
<LinearLayout
    android:layout_width="match_parent"
    android:layout_height="wrap_content"
    android:orientation="horizontal">
    <TextView
        android:id="@+id/textView3"
        android:layout_width="0dp"
        android:layout_height="wrap_content"
        android:layout_weight="1"
        android:gravity="center"
        android:textSize="30sp"
        android:text="成绩" />
    <EditText
        android:id="@+id/edit_score"
        android:layout_width="0dp"
        android:layout_height="wrap_content"
        android:layout_weight="3"
        android:textSize="30sp"
        android:inputType="number" />
</LinearLayout>
<LinearLayout
    android:layout_width="match_parent"
    android:layout_height="wrap_content"
    android:orientation="horizontal">
    <Button
        android:id="@+id/btn_insert"
        android:layout_width="wrap_content"
        android:layout_height="wrap_content"
        android:layout_weight="1"
        android:textSize="30sp"
        android:layout_marginRight="20sp"
        android:text="新增" />
    <Button
        android:id="@+id/btn_query"
        android:layout_width="wrap_content"
        android:layout_height="wrap_content"
```

```xml
            android:layout_weight="1"
            android:textSize="30sp"
            android:text="查询" />
    </LinearLayout>
    <LinearLayout
        android:layout_width="match_parent"
        android:layout_height="wrap_content"
        android:orientation="horizontal">
        <Button
            android:id="@+id/btn_del"
            android:layout_width="wrap_content"
            android:layout_height="wrap_content"
            android:layout_weight="1"
            android:layout_marginRight="20sp"
            android:text="删除"
            android:textSize="30sp" />
        <Button
            android:id="@+id/btn_update"
            android:layout_width="wrap_content"
            android:layout_height="wrap_content"
            android:layout_weight="1"
            android:text="修改"
            android:textSize="30sp" />
    </LinearLayout>
    <EditText
        android:id="@+id/edit_show"
        android:layout_width="365dp"
        android:layout_height="300dp"
        android:gravity="top"
        android:ems="10"
        android:inputType="textMultiLine"
        android:text=""
        android:textSize="30sp" />
</LinearLayout>
```

（2）创建数据库工具类 DBUtil。

定义一个 DBUtil 类，用于实现对数据库信息的映射，DBUtil.java 代码如下：

```java
public class DBUtil {
    //数据库名称和表名称
    public static final String DB_NAME="student.db";
    public static final String TABLE_NAME="studentScore";
    //数据库版本号
    public static final Integer VERSION = 1;
    //表中各字段的名称
    public static final String STU_ID="_id";
    public static final String STU_NAME="name";
    public static final String STU_SCORE="score";
    //创建表的 SQL 语句
    public static final String CREATE_TABLE="create table   "+
            TABLE_NAME+
            " ( "+ STU_ID+"   integer primary key autoincrement，"
            + STU_NAME+"   text not null，"
            +STU_SCORE+"   integer );";
}
```

(3)创建数据库辅助操作类 MyDBHelper。

创建一个数据库辅助操作类 MyDBHelper,继承 SQLiteOpenHelper 类,重写 onCreat()方法和 onUpgrade()方法。MyDBHelper.java 代码如下:

```java
public class MyDBHelper extends SQLiteOpenHelper {
    public MyDBHelper(Context context, String name,
                     SQLiteDatabase.CursorFactory factory, int version) {
        super(context, name, factory, version);
    }
    public MyDBHelper(Context context){
        super(context,DBUtil.DB_NAME,null,DBUtil.VERSION);
    }
    @Override
    public void onCreate(SQLiteDatabase db) {
        db.execSQL(DBUtil.CREATE_TABLE);   //创建数据表
    }
    @Override
    public void onUpgrade(SQLiteDatabase db, int oldVersion, int newVersion) {
        db.execSQL("DROP TABLE IF EXISTS "+DBUtil.TABLE_NAME);
        onCreate(db);
    }
}
```

(4)编写 MainActivity。

在 MainActivity 中编写界面上各个按钮的单击事件处理程序,实现增、删、改、查的功能,执行删和改操作时需输入的学号。MainActivity.java 代码如下:

```java
public class MainActivity extends AppCompatActivity
                         implements View.OnClickListener {
    private EditText edit_num;
    private EditText edit_name;
    private EditText edit_score;
    private Button btn_insert;
    private Button btn_del;
    private Button btn_query;
    private Button btn_update;
    private EditText edit_show;
    MyDBHelper dbHelper;
    SQLiteDatabase db;
    ContentValues values;
    @Override
    protected void onCreate(Bundle savedInstanceState) {
        super.onCreate(savedInstanceState);
        setContentView(R.layout.activity_main);
        initView();
        dbHelper=new MyDBHelper(this);
    }
    private void initView() {
        edit_num = (EditText) findViewById(R.id.edit_num);
        edit_name = (EditText) findViewById(R.id.edit_name);
        edit_score = (EditText) findViewById(R.id.edit_score);
        btn_insert = (Button) findViewById(R.id.btn_insert);
        btn_del = (Button) findViewById(R.id.btn_del);
```

```java
        btn_query = (Button) findViewById(R.id.btn_query);
        btn_update = (Button) findViewById(R.id.btn_update);
        edit_show = (EditText) findViewById(R.id.edit_show);
        btn_insert.setOnClickListener(this);
        btn_del.setOnClickListener(this);
        btn_query.setOnClickListener(this);
        btn_update.setOnClickListener(this);
    }

    @Override
    public void onClick(View v) {
        String num=edit_num.getText().toString();
        String name=edit_name.getText().toString();
        String score=edit_score.getText().toString();
        switch (v.getId()) {
            case R.id.btn_insert:
                if (num.isEmpty()) {
                    Toast.makeText(this,"学号为空！", Toast.LENGTH_SHORT).show();
                    return;
                }
                if (name.isEmpty()) {
                    Toast.makeText(this,"姓名为空！", Toast.LENGTH_SHORT).show();
                    return;
                }
                if (score.isEmpty()) {
                    Toast.makeText(this,"成绩为空！", Toast.LENGTH_SHORT).show();
                    return;
                }
                db=dbHelper.getWritableDatabase();
                values=new ContentValues();
                values.put(DBUtil.STU_ID, Integer.valueOf(num));
                values.put(DBUtil.STU_NAME, name);
                values.put(DBUtil.STU_SCORE,Integer.valueOf(score));
                db.insert(DBUtil.TABLE_NAME, null, values);
                db.close();
                //执行插入操作后，清除编辑框中的内容
                edit_num.setText("");
                edit_name.setText("");
                edit_score.setText("");
                break;
            case R.id.btn_query:
                edit_show.setText("");
                db=dbHelper.getWritableDatabase();
                Cursor cursor=db.query(DBUtil.TABLE_NAME, null, null, null, null, null, null);
                if (cursor==null){
                    edit_show.append("无结果");
                }else{
                    //遍历查询结果集，并显示在界面的编辑框中
                    while (cursor.moveToNext()){
                        String id=cursor.getString(0);
                        String stuname=cursor.getString(1);
                        String stuscore=cursor.getString(2);
                        edit_show.append(id+"    "+stuname+"    "+ stuscore+"\n");
                    }
```

```
                    }
                    db.close();
                    break;
                case R.id.btn_del:
                    db=dbHelper.getWritableDatabase();
                    //按学号删除
                    if (num.isEmpty()){
                        Toast.makeText(this,"请输入学号！", Toast.LENGTH_SHORT).show();
                        return;
                    }
                    int count=db.delete(DBUtil.TABLE_NAME, DBUtil.STU_ID+"=?", new String[]{num});
                    if (count>0){
                        edit_show.setText("删除"+count+"条记录!");
                    }else {
                        Toast.makeText(this,"无此学号！", Toast.LENGTH_SHORT).show();
                    }
                    db.close();
                    break;
                case R.id.btn_update:
                    if (num.isEmpty()){
                        Toast.makeText(this,"请输入学号！", Toast.LENGTH_SHORT).show();
                        return;
                    }
                    db=dbHelper.getWritableDatabase();
                    values=new ContentValues();
                    values.put(DBUtil.STU_SCORE, Integer.valueOf(score));
                    int i=db.update(DBUtil.TABLE_NAME, values,
                            DBUtil.STU_ID+"=?",new String[]{num});
                    if (i>0){
                        Toast.makeText(this,i+"条数据被修改！", Toast.LENGTH_SHORT).show();
                    }else {
                        Toast.makeText(this,"无此学号！", Toast.LENGTH_SHORT).show();
                    }
                    db.close();
                    break;
            }
        }
    }
```

运行程序，在图 5.11 的操作界面上输入学生信息，如图 5.12 所示。单击"新增"按钮，向数据表中插入 1 条学生记录。重复上述操作，再添加 1 条记录，然后单击"查询"按钮，查询结果如图 5.13 所示。

执行修改操作，将学号为 1001 的学生成绩修改为 95 分，然后单击"查询"按钮，查询结果如图 5.14 所示。

输入学号 1001，单击"删除"按钮执行操作，界面上显示"删除 1 条记录"，结果如图 5.15 所示。然后单击"查询"按钮，查询结果如图 5.16 所示，数据表中的学号为 1001 的学生信息被删除了。

本例演示了 SQLite 数据库的使用方法，即将学生信息存储在本地 SQLite 数据库中。在实际 Android 开发中，除使用以上介绍的本地数据存储方式外，还可以采用其他方式，如通过网络传输将数据存储到远程机器上。在实际编程时，需要根据程序的功能及其具体需要来决定采用哪种方式存储数据，或者采用哪些方式的组合更为适合。

图 5.12　插入数据　　　图 5.13　新增后查询　　　图 5.14　修改成绩

图 5.15　删除　　　图 5.16　删除后查询结果

习题 5

1．完成习题 4 中考试系统的记住密码功能。用户第一次登录时如果选择记住密码，则登录时将账号和密码保存到 SharedPreferences 中，下次启动程序时登录界面中自动填写账号和密码。

2．修改该考试系统的记住密码功能，登录时将账号和密码保存到内部文件中。

3．完成该考试系统的数据存储功能。当单击用户注册界面的"注册"按钮时，将注册信息存储到 SQLite 数据库中。

第 6 章　内容提供者

Android 应用程序运行在不同的进程空间中，应用程序间的数据是不能够直接访问的。但在 Android 开发中有时也需要访问其他应用程序的数据，为此 Android 提供了 ContentProvider（内容提供者）组件，专门用于在不同应用程序之间实现数据共享。本章将主要介绍如何使用 ContentProvider 实现数据的共享，以及使用 ContentProvider 访问系统数据。

6.1　ContentProvider 简介

在 Android 系统的手机中，ContentProvider 最典型的应用是在短信和联系人之间共享数据。例如，当发送一条短信时，需要用到联系人的相关信息，此时短信应用就会通过 ContentProvider 提供的接口访问 Android 系统中的联系人信息，并从中选择联系人。Android 系统中有很多内置数据，如音频、视频、图像等，都是通过 ContentProvider 提供给用户使用的。

ContentProvider 是应用程序之间共享数据的一种接口机制，提供了一套标准 API 用来获取和操作数据。通过 ContentProvider，应用程序可以指定需要共享的数据，而其他应用程序可以在不知道数据来源、路径的情况下，对共享数据进行查、增、删和改的操作。

ContentProvider 底层数据源可以是 SQLite 数据库、文件系统、SharedPreferences 和网络数据等存储方式，ContentProvider 提供了对底层数据存储方式的抽象，为应用程序间的数据交互提供了一个安全的环境。它准许应用程序把自己的数据根据需求开放给其他应用程序进行增、删、改、查的操作，而不用担心直接开放数据库权限而带来的安全问题。

一个应用程序通过 ContentProvider 将自己的数据共享给其他应用程序，其他应用程序通过 ContentResolver 对象来访问共享的数据。ContentProvider 以 URI（自身数据集的唯一标识）的形式对外提供数据，ContenrResolver 根据 URI 来访问数据。ContentProvider 的工作原理如图 6.1 所示。

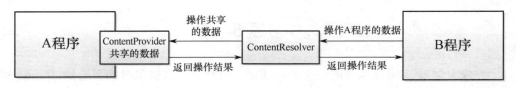

图 6.1　ContentProvider 的工作原理

6.2　URI 简介

URI（Uniform Resource Identifier，统一资源标识符）是一个用于标识某个互联网资源名称的字符串。该标识允许用户对任何（包括本地和互联网）的资源通过特定的协议进行交互操作。

在 ContentProvider 机制中，每个 ContentProvider 必须提供一个对外统一的 URI 标识其数据集，ContentResolver 对象通过 URI 定位 ContentProvider 提供的资源。

ContentProvider 使用的 URI 语法格式如下：

```
content://<authority>/<data_path>/<id>
```

其中的参数含义如下。

content://：通用前缀，表示该 URI 用于 ContentProvider 定位资源。

\<authority\>：授权者名称，一般由类的小写全称组成，用于标识唯一的 ContentProvider。

\<data_path\>：数据路径，用于确定数据集。ContentProvider 仅提供一个数据集，数据路径可以省略；如果提供多个数据集，则必须用数据路径指明具体是哪一个数据集。

\<id\>：数据编号，用于确定数据集中的一条唯一记录，匹配数据集中的 _ID 字段的值。如果请求的数据不仅限于一条，则\<id\>可以省略。

例如，请求整个 student 数据集的 URI 应写为：

```
content://com.example.studentprovider/student
```

而请求 student 数据集中第三条数据的 URI，则应写为：

```
content://com.example.studentprovider/student/3
```

ContentProvider 中的数据集类似于数据库中的数据表，每行是一条记录，每列具有相同的数据类型。ContentProvider 可以提供多个数据集，每个数据集分配一个独立且唯一的 URI，调用者可使用 URI 对不同数据集的数据进行操作。

Android 系统提供了两个用于操作 URI 的工具类，分别为 UriMatcher 和 ContentUris。

UriMatcher 类用于对 ContentProvider 中的 URI 进行匹配。该工具类提供了以下两种方法。

（1）void addURI(String authority,String path,int code)

该方法的第一个参数表示 URI 的 authority 部分，第二个参数表示 URI 的 path 部分，第三个参数表示 URI 匹配成功后返回的匹配码。

（2）int match(Uri uri)

该方法对指定的 URI 进行匹配，如果匹配成功，则返回匹配码，否则返回常量 UriMatcher.NO_MATCH。

使用 UriMatcher 类对 URI 进行匹配时，首先初始化一个 UriMatcher 实例，然后使用该实例的 addURI()方法将 URI 注册到 UriMatcher 中，代码如下：

```
UriMatcher uriMatcher = new UriMatcher(UriMatcher.NO_MATCH);
uriMatcher.addURI("com.example.studentprovider", "student", STUDENT);
uriMatcher.addURI("com.example.studentprovider", " student/#", STUDENT_ID);
```

上述代码中，常量 UriMatcher.NO_MATCH 表示不匹配任何路径的返回码(-1)。

为了便于使用 URI，可将 URI 的授权者名称和数据路径等声明为静态常量。声明 RUI 的代码如下：

```
private static final String AUTHORITY = "com.example.studentprovider";
private static final String PATH _SINGLE = "student/#";
private static final String PATH _MULTIPLE = "student";
private static final String CONTENT_URI_STRING = "content://"+ AUTHORITY+ " /"+ PATH _MULTIPLE;
```

```java
private static final Uri CONTENT_URI = Uri.parse(CONTENT_URI_STRING);
private static final int MULTIPLE_STUDENT = 1;    //多条数据的返回码
private static final int SINGLE_STUDENT = 2;    //单条数据的返回码
private static final UriMatcher uriMatcher;
static {
    uriMatcher = new UriMatcher(UriMatcher.NO_MATCH);
    uriMatcher.addURI(AUTHORITY, PATH_SINGLE, SINGLE_STUDENT);
    uriMatcher.addURI(AUTHORITY, PATH_MULTIPLE, MULTIPLE_STUDENT);
}
```

使用 uriMatcher.match(uri)方法对指定的 URI 进行匹配的代码如下:

```java
int uriType = uriMatcher.match(uri);
switch (uriType){
    case    MULTIPLE_STUDENT:
        //多条数据的处理过程
         break;
    case SINGLE_STUDENT:
        //单条数据的处理过程
        break;
    default:
        throw new UnsupportedOperationException("Not yet implemented");
}
```

ContentUris 工具类用于操作 URI 字符串,该工具类提供了以下两种方法:
- withAppendedId(uri,id)用于为 URI 路径加上 ID 部分;
- parseId(uri)用于从指定的 URI 中解析出所包含的 ID 值。

6.3 开发 ContentProvider

一个应用程序如果要将自己的数据共享给其他应用程序,就需要先定义自己的 ContentProvider 子类,然后在 AndroidManifest.xml 中注册,其他应用程序可以获取 ContentResolver 对象通过 URI 访问数据。

6.3.1 创建和注册 ContentProvider

1. 继承 ContentProvider 类创建一个子类

在应用程序包名处单击右键,并在弹出的菜单中执行"New"→"Other"→"Content Provider"菜单命令,弹出"New Android Component"对话框,如图 6.2 所示。

输入内容提供者的 Class Name(名称)和 URI Authorities(唯一标识,通常使用包名),单击"Finish"按钮完成创建。

2. 实现相关的方法

新创建的类继承 ContentProvider 后,需要重载 onCreate( )方法、insert( )方法、delete( )方法、update( )方法、query( )方法、getType( )方法和 getContext( )方法,各方法的语法格式如表 6.1 所示。

第 6 章　内容提供者

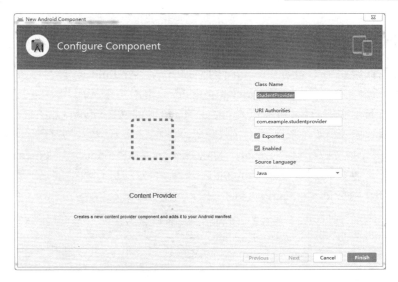

图 6.2　创建 ContentProvider

表 6.1　ContentProvider 的常用方法

方　　法	功　能　描　述
public abstract boolean onCreate( )	创建 ContentProvider 后会被调用
public abstract Uri insert(Uri uri,ContentValues values)	根据 URI 插入 values 对应的数据
public abstract int delete(Uri uri,String selection,String[] selectionArgs)	根据 URI 删除 selection 条件所匹配的全部记录
public abstract int update(Uri uri,ContentValues values,String selection, String[] selectionArgs)	根据 URI 修改 selection 条件所匹配的全部记录
public abstract Cursor query(Uri uri,String[] projection,String selection, String[] selectionArgs,String sortOrder)	根据 URI 查询 selection 条件所匹配的全部记录
public abstract String getType(Uri uri)	获得当前 URI 所代表的 MIME 数据类型
public final Context getContext()	获得 Context 对象

3．在 AndroidManifest.XML 中注册 ContentProvider

创建 ContentProvider 子类后必须在应用程序的 AndroidManifest.xml 中进行注册，否则该 ContentProvider 子类对于 Android 系统是不可见的。StudentProvider 的注册代码如下：

```
<provider
    android:name=".StudentProvider"
    android:authorities="com.example.studentprovider"
    android:enabled="true"
    android:exported="true">
</provider>
```

其中参数的属性如下。

name 属性用来创建 StudentProvider 的类名全称（包名可以省略）。

authorities 属性用来标识一个唯一的 ContentProvider。

enabled 属性表示创建的 ContentProvider 能否被系统实例化。

exported 属性表示是否支持其他应用调用当前组件，true 表示支持，false 表示不支持。

还可以加入 android:readPermission 属性或 android: writePermission 属性来控制其权限。

| 163

6.3.2 使用 ContentResolver 操作数据

调用者不能直接调用 ContentProvider 的接口方法，Android 提供了 ContentResolver（内容解析者），可通过 URI 间接调用 ContentProvider。由于每个应用程序的上下文都有一个默认的 ContentResolver 实例对象，可通过 Context 的 getContentResolver()方法来获取该实例对象。

ContentResolver 提供一系列增、删、改、查的方法对数据进行操作，这些方法与 ContentProvider 中的方法是一一对应的。实际上，ContentProvider 与 ContentResolver 就是通过 URI 进行数据交换，调用者使用 ContentResolver 的方法进行数据的增、删、改、查操作，用于调用指定 URI 对应的 ContentProvider 中的方法，如表 6.2 所示。

表 6.2　ContentResolver 的方法

方　　法	功　能　描　述
insert(Uri uri,ContentValues values)	向 URI 对应的 ContentProvide 中增加 values 匹配的数据
delete(Uri uri,String where,String[] selectionArgs)	删除 URI 对应的 ContentProvide 中 where 匹配的数据
update(Uri uri,ContentValues values,String where, String[] selectionArgs)	更新 URI 对应的 ContentProvide 中 where 匹配的数据
query(Uri uri,String[] projection,String selection, String[],selectionArgs, String sortOder)	查询 URI 对应的 ContentProvide 中 where 匹配的数据

一般来说，每个 ContentProvider 都只有一个实例（单例模式），但可以与多个在不同应用或进程中的 ContentResolver 进行交互。也就是说，当多个应用程序通过 ContentResolver 来操作 ContentProvider 提供的数据时，ContentResolver 调用的数据操作将会委托给同一个 ContentProvider 处理。

【例 6.1】使用 ContentProvider 实现数据共享。

在【例 5.4】中使用 SQLite 数据库存储学生成绩信息（studentScore 数据表），本例将使用 ContentProvider 实现 studentScore 数据表中数据的共享，并另外开发一个应用程序来访问共享数据。

（1）修改【例 5.4】中 DBUtil 类，添加 URI 相关信息。

DBUtil.java 代码如下：

```java
public class DBUtil {
    public static final String DB_NAME="student.db";
    public static final String TABLE_NAME="studentScore";
    //表中各字段
    public static final String STU_ID="_id";
    public static final String STU_NAME="name";
    public static final String STU_SCORE="score";
    //创建表的 SQL 语句
    public static  final String CREATE_TABLE="create table"+"   "+TABLE_NAME+
            " ( "+ STU_ID+"+"   "+integer primary key autoincrement,"
            + STU_NAME+"+"   "+text not null,"
            +STU_SCORE+"+"   "+integer );";
    //URI 相关信息
    public static final String AUTHORITY="com.example.sqlite";
```

```
        public static final String    PATH_SINGLE="studentScore/#";
        public   static   final    String PATH_MULTIPLE="studentScore";
        public static final  String  CONTENT_URI_STRING="content://"+AUTHORITY+"/"+DBUtil.PATH_ MULTIPLE;
        public   static    final Uri CONTENT_URI=Uri.parse(CONTENT_URI_STRING);
    }
```

加粗部分的代码为增加的 URI 相关信息。

（2）创建 StudentProvider 类。

在【例 5.4】工程中创建 StudentProvider 类，提供对学生成绩信息的插入、查询、更新和删除功能。StudentProvider 在 AndroidManifest.xml 中的注册代码如下：

```
<provider
    android:name=".StudentProvider"
    android:authorities="com.example.sqlite"
      android:enabled="true"
      android:exported="true">
</provider>
```

StudentProvider.java 的代码如下：

```java
public class StudentProvider extends ContentProvider {
    private SQLiteDatabase db;
    private MyDBHelper dbHelper;
    private static final int MULTIPLE_STUDENT=1;
    private static final int SINGLE_STUDENT=2;
    private static final UriMatcher uriMatcher;
    static {
        uriMatcher=new UriMatcher(UriMatcher.NO_MATCH);
        uriMatcher.addURI(DBUtil.AUTHORITY, DBUtil.PATH_MULTIPLE, MULTIPLE_STUDENT);
        uriMatcher.addURI(DBUtil.AUTHORITY, DBUtil.PATH_SINGLE, SINGLE_STUDENT);
    }
    @Override
    public boolean onCreate() {
        Context context=getContext();
        dbHelper=new MyDBHelper(context);
        db=dbHelper.getWritableDatabase();
        if (db==null){
            return false;
        }else
            return true;
    }
    @Override
    public int delete(Uri uri, String selection, String[] selectionArgs) {
        int count=0;
        switch (uriMatcher.match(uri)){
            case MULTIPLE_STUDENT:
                count=db.delete(DBUtil.TABLE_NAME,selection,selectionArgs);
                break;
            case SINGLE_STUDENT:
                count=db.delete(DBUtil.TABLE_NAME, DBUtil.STU_ID+"=?", selectionArgs);
            default:
                throw new UnsupportedOperationException("Not yet implemented");
        }
        return count;
```

```java
        }
        @Override
        public Uri insert(Uri uri, ContentValues values) {
            switch (uriMatcher.match(uri)){
                case MULTIPLE_STUDENT:
                    long id=db.insert(DBUtil.TABLE_NAME, null, values);
                    if (id>0){
                        Uri newUri= ContentUris.withAppendedId(DBUtil.CONTENT_URI, id);
                        return newUri;
                    }
                    break;
                default:
                    throw new UnsupportedOperationException("Not yet implemented");
            }
            return null;
        }
        @Override
        public Cursor query(Uri uri, String[] projection, String selection, String[] selectionArgs,
                            String sortOrder) {
            Cursor cursor;
            switch (uriMatcher.match(uri)){
                case MULTIPLE_STUDENT:
                    cursor=db.query(DBUtil.TABLE_NAME, projection, selection, selectionArgs,
                                    null, null, sortOrder);
                    break;
                case SINGLE_STUDENT:
                    cursor=db.query(DBUtil.TABLE_NAME, projection, DBUtil.STU_ID+"=",
                                    selectionArgs, null, null, sortOrder);
                    break;
                default:
                    throw new UnsupportedOperationException("Not yet implemented");
            }
            return cursor;
        }
        @Nullable
        @Override
        public String getType(@NonNull Uri uri) {
            return null;
        }
        @Override
        public int update(Uri uri, ContentValues values, String selection, String[] selectionArgs) {
            int count;
            switch (uriMatcher.match(uri)){
                case MULTIPLE_STUDENT:
                    count=db.update(DBUtil.TABLE_NAME, values, selection, selectionArgs);
                    break;
                case SINGLE_STUDENT:
                    count=db.update(DBUtil.TABLE_NAME, values, DBUtil.STU_ID+"=?",
                                    selectionArgs);
                    break;
                default:
                    throw new UnsupportedOperationException("Not yet implemented");
            }
```

```
            return count;
        }
    }
```

(3) 创建应用程序 Resolver。

创建一个应用程序 Resolver，通过 ContentResolver 访问【例 5.4】StudentProvider 已提供的学生成绩数据。该应用程序会用到 DBUtil 类中的数据，需要将该类复制到 Resolver 应用程序相应的目录中。该应用程序的主界面与【例 5.4】相同，此处不再重复。

应用程序 Resolver 的 MainActivity.Java 代码如下：

```
public class MainActivity extends AppCompatActivity implements View.OnClickListener {
    private ContentResolver resolver;
    private Uri uri;
    private ContentValues values;
    private EditText edit_num;
    private EditText edit_name;
    private EditText edit_score;
    private Button btn_insert;
    private Button btn_del;
    private Button btn_query;
    private Button btn_update;
    private EditText edit_show;
    @Override
    protected void onCreate(Bundle savedInstanceState) {
        super.onCreate(savedInstanceState);
        setContentView(R.layout.activity_main);
        initView();
    }
    private void initView() {
        edit_num = (EditText) findViewById(R.id.edit_num);
        edit_name = (EditText) findViewById(R.id.edit_name);
        edit_score = (EditText) findViewById(R.id.edit_score);
        btn_insert = (Button) findViewById(R.id.btn_insert);
        btn_del = (Button) findViewById(R.id.btn_del);
        btn_query = (Button) findViewById(R.id.btn_query);
        btn_update = (Button) findViewById(R.id.btn_update);
        edit_show = (EditText) findViewById(R.id.edit_show);
        btn_insert.setOnClickListener(this);
        btn_del.setOnClickListener(this);
        btn_query.setOnClickListener(this);
        btn_update.setOnClickListener(this);
    }
    @Override
    public void onClick(View v) {
        resolver=getContentResolver();
        uri= DBUtil.CONTENT_URI;
        switch (v.getId()) {
            case R.id.btn_insert:
                values=new ContentValues();
                values.put(DBUtil.STU_ID,edit_num.getText().toString());
                values.put(DBUtil.STU_NAME,edit_name.getText().toString());
                values.put(DBUtil.STU_SCORE,Integer.valueOf(edit_score.getText().toString()));
                resolver.insert(uri,values);
```

```
            break;
        case R.id.btn_del:
            resolver.delete(uri,null,null);
            break;
        case R.id.btn_query:
            edit_show.setText("");
            Cursor cursor= resolver.query(uri,null,null,null,null,null);
            if (cursor.getCount()>0){
                while (cursor.moveToNext()){
                    edit_show.append(cursor.getString(0)+" ");
                    edit_show.append(cursor.getString(1)+" ");
                    edit_show.append(cursor.getString(2)+"\n");
                }
            }
            break;
        case R.id.btn_update:
            values=new ContentValues();
            values.put(DBUtil.STU_SCORE,Integer.valueOf(edit_score.getText().toString()));
            resolver.update(uri,values,DBUtil.STU_ID+"=?",new
                            String[]{edit_num.getText().toString()});
            break;
    }
}
```

按如下步骤进行操作。

（1）运行【例 5.4】的应用程序，单击"查询"按钮，查看 SQLite 数据库中的数据，结果如图 6.3 所示。

（2）运行本例的应用程序 Resolver，单击"查询"按钮，查询结果如图 6.4 所示，与 SQLite 数据库中的数据一致。

图 6.3　SQLite 数据库中的信息　　图 6.4　resolver 查询结果

（3）通过应用程序 Resolver 输入一个学生的信息，单击"新增"按钮，然后单击"查询"按钮，查询结果如图 6.5 所示。

（4）单击【例 5.4】应用程序界面上的"查询"按钮，查询结果如图 6.6 所示，与应用程序 Resolver 查询结果一致。

图 6.5 resolver 添加数据　　　图 6.6 SQLite 查询结果

删除和修改功能不再演示，感兴趣的读者可自行测试。

6.4 监听 ContentProvider 的数据改变

当 ContentProvider 将数据共享出来后，ContentResolver 会根据业务需求对数据进行操作，在有些时候，应用程序需要实时监听 ContentProvider 中数据的变化，继而做一些相应的处理。为此，Android 提供了抽象类 ContentObserver（内容观察者）。

为了监听 ContentProvider 的数据变化，需要通过 ContentResolver 向指定的 URI 注册 ContentObserver 监听器。ContentResolver 提供了如下方法来注册监听器：

public final void registerContentObserver(Uri uri, Boolean notifyForDescendents, ContentObserver observer)

该方法为指定的 URI 注册一个 ContentObserver 派生类实例，当指定的 URI 发生改变时，回调该实例对象去处理。参数 URI 是该监听器所监听 ContentProvider 的 URI。参数 notifyForDescendents 为 false 时表示精确匹配，即只匹配该 URI；为 true 时表示可以同时匹配其派生的 URI。参数 Observer 即该监听器的实例。

要使用 ContentObserver 观察数据变化，必须在 ContentProvider 的 delete()方法、insert()方法、update()方法中调用 ContentResolver 的 notifyChange()方法。当 ContentObserver 观察到指定 URI 代表的数据发生变化时，就会触发 ContentObserver 的 onChange()方法，在该方法中使用 ContentResolver 可以查询到变化的数据。ContentProvider、ContentResolver 和 ContentObserver 三者之间的关系如图 6.7 所示。

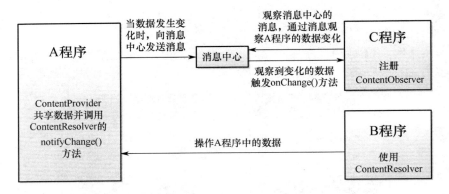

图 6.7 ContentProvider、ContentResolver 和 ContentObserver 之间的关系

【例 6.2】使用 ContentObserver 监听数据变化。

创建一个应用程序 Observer，使用 ContentObserver 监听【例 5.4】中 StudentProvider 提供的数据变化。应用程序 Observer 只需要监听对数据的操作，不需要创建主界面，使用默认界面即可。本例通过在 MainActivity 中注册 ContentObserver，监听 StudentProvider 提供的数据是否发生变化。

（1）MainActivity 中注册 ContentObserver 监听器。

首先，在 MainActivity 中创建一个内部类 MyObserver，继承 ContentObserver 类，重写 onChange()方法，当监听的数据发生变化时输出日志信息"数据变化了"。

然后，在 MainActivity 的 onCreate()方法中注册该监听器，在 onDestroy()方法中注销该监听器。

```
public class MainActivity extends AppCompatActivity {
    @Override
    protected void onCreate(Bundle savedInstanceState) {
        super.onCreate(savedInstanceState);
        setContentView(R.layout.activity_main);
        Uri uri= Uri.parse("content://com.example.sqlite/studentScore");
        getContentResolver().registerContentObserver(uri, true, new MyObserver(new Handler()));
    }
    @Override
    protected void onDestroy() {
        super.onDestroy();
        getContentResolver().unregisterContentObserver(new MyObserver(new Handler()));
    }
    private class   MyObserver extends ContentObserver{
        public MyObserver(Handler handler) {
            super(handler);
        }
        @Override
        public void onChange(boolean selfChange) {
            super.onChange(selfChange);
            Log.i("TAG", "数据变化了");
        }
```

        }
    }

（2）修改 StudentProvider 类。

修改【例 5.4】中 StudentProvider 类的 delete()方法、insert()方法、update()方法，调用 ContentResolver 的 notifyChange()方法，当 StudentProvider 提供的数据发生变化时，就会触发 MyObserver 的 onChange()方法。以下代码中加粗部分是添加的代码。

```
public int delete(Uri uri, String selection, String[] selectionArgs) {
    int count=0;
    switch (uriMatcher.match(uri)){
        case MULTIPLE_STUDENT:
            count=db.delete(DBUtil.TABLE_NAME,selection,selectionArgs);
            break;
        case SINGLE_STUDENT:
            count=db.delete(DBUtil.TABLE_NAME, DBUtil.STU_ID+"=?", selectionArgs);
        default:
            throw new UnsupportedOperationException("Not yet implemented");
    }
    if (count>0){
        getContext().getContentResolver().notifyChange(uri, null);
    }
    return count;
}
public Uri insert(Uri uri, ContentValues values) {
    switch (uriMatcher.match(uri)){
        case MULTIPLE_STUDENT:
            long id=db.insert(DBUtil.TABLE_NAME, null, values);
            if (id>0){
                Uri newUri= ContentUris.withAppendedId(DBUtil.CONTENT_URI, id);
                getContext().getContentResolver().notifyChange(newUri, null);
                return newUri;
            }
            break;
        default:
            throw new UnsupportedOperationException("Not yet implemented");
    }
    return null;
}
public int update(Uri uri, ContentValues values, String selection, String[] selectionArgs) {
    int count;
    switch (uriMatcher.match(uri)){
        case MULTIPLE_STUDENT:
            count=db.update(DBUtil.TABLE_NAME, values, selection, selectionArgs);
            break;
        case SINGLE_STUDENT:
```

```
                    count=db.update(DBUtil.TABLE_NAME, values, DBUtil.STU_ID+"=?", selectionArgs);
                    break;
                default:
                    throw new UnsupportedOperationException("Not yet implemented");
        }
        if (count>0){
            getContext().getContentResolver().notifyChange(uri, null);
        }
        return count;
    }
```

代码修改后需要重新运行【例 5.4】的应用程序，然后运行应用程序 Observer，打开应用程序 Resolver，添加一个学生的信息，LogCat 窗口的日志信息如图 6.8 所示，应用程序 Observer 监测到了 StudentProvider 中数据的变化。

```
logcat
05-03 14:25:03.493 I/TAG: 数据变化了
```

图 6.8　监听数据变化

6.5　使用系统内置的 ContentProvider

Android 系统本身提供了大量的 ContentProvider，程序员可以自己开发应用程序调用这些 ContentProvider 访问系统数据，如联系人信息、多媒体信息和短信等。

Android 系统提供了以下几个 URI 来访问联系人信息。

（1）ContactsContract.Contacts.CONTENT_URI：联系人的 URI。

（2）ContactsContract.CommonDataKinds.Phone.CONTENT_URI：联系人电话的 URI。

（3）ContactsContract.CommonDataKinds.Email.CONTENT_URI：联系人 E-mail 的 URI。

Android 系统提供访问多媒体内容的 URI 如下。

（1）MediaStore.Audio.Media.EXTERNAL_CONTENT_URI：存储在外部存储器（SD 卡）上音频文件内容的 ContentProvider 的 URI。

（2）MediaStore.Audio.Media.INTERNAL_CONTENT_URI：存储在手机内部存储器（SD 卡）上音频文件内容的 ContentProvider 的 URI。

（3）MediaStore.Images.Media.EXTERNAL_CONTENT_URI：存储在外部存储器（SD 卡）上图片文件内容的 ContentProvider 的 URI。

（4）MediaStore.Images.Media.INTERNAL_CONTENT_URI：存储在手机内部存储器（SD 卡）上图片文件内容的 ContentProvider 的 URI。

（5）MediaStore.Video.Media.EXTERNAL_CONTENT_URI：存储在外部存储器（SD 卡）上视频文件内容的 ContentProvider 的 URI。

（6）MediaStore.Video.Media.INTERNAL_CONTENT_URI：存储在手机内部存储器（SD 卡）上视频文件内容的 ContentProvider 的 URI。

下面通过一个获取手机联系人的案例来演示如何调用系统内置的 ContentProvider 来访问系统数据。

【例 6.3】获取手机联系人。

Android 中联系人的信息是通过 ContentProvider 提供给外部应用使用的，使用时只需根据系统联系人 ContentProvider 的 URI 即可获取所需数据。本例中要获取系统中联系人的名字和所有电话号码，并在 ListView 列表中显示出来。要访问系统联系人信息，应用程序需要有"android.permission.READ_CONTACTS"权限。

（1）界面设计。

本例需要在 ListView 控件中显示联系人信息，因此应用程序的主界面比较简单，只有一个 ListView 控件。每个列表项都会显示联系人的姓名和电话号码。

在 res/layout 文件夹中创建一个 Item 界面的布局文件 list.xml，在该文件中放置两个 TextView 控件，分别显示联系人姓名和电话号码。布局文件代码如下：

```xml
<?xml version="1.0" encoding="utf-8"?>
<LinearLayout xmlns:android="http://schemas.android.com/apk/res/android"
    android:orientation="vertical"
    android:layout_width="match_parent"
    android:layout_height="match_parent">
    <TextView
        android:id="@+id/tv_name"
        android:layout_width="wrap_content"
        android:layout_height="wrap_content"
        android:textSize="20sp"
        android:text="联系人姓名" />
    <TextView
        android:id="@+id/tv_tel"
        android:layout_width="wrap_content"
        android:layout_height="wrap_content"
        android:textSize="15sp"
        android:inputType="textMultiLine"
        android:text="电话号码"  />
</LinearLayout>
```

（2）编写 MainActivity。

```java
public class MainActivity extends AppCompatActivity    {
    private ListView listView;
    private ArrayList<Map<String, Object>> contactsInfo;
    private Map<String, Object> map;
    SimpleAdapter adapter;
    @Override
    protected void onCreate(Bundle savedInstanceState) {
        super.onCreate(savedInstanceState);
        setContentView(R.layout.activity_main);
        initView();
        requestPermission();
    }
    public void queryContact() {
        String contactId,name,number;
        ContentResolver resolver = getContentResolver();
        //查询联系人列表，但不显示号码
```

```java
            Cursor cursor = resolver.query(ContactsContract.Contacts.CONTENT_URI,
                        null, null, null, null);
            contactsInfo=new ArrayList<Map<String, Object>>();
            //遍历查询结果，获取系统中所有联系人
            while (cursor.moveToNext()) {
                map = new HashMap<String, Object>();
                //获取联系人 ID
                contactId = cursor.getString(cursor.getColumnIndex(ContactsContract.Contacts._ID));
                //获取联系人姓名
                name = cursor.getString(cursor.getColumnIndex(ContactsContract.
                            Contacts.DISPLAY_NAME));
                //根据 ID 查找该联系人的电话号码
                Cursor numbers = resolver.query(ContactsContract.CommonDataKinds.Phone.
                            CONTENT_URI,null,
                        ContactsContract.CommonDataKinds.Phone.CONTACT_ID+"=?",
                        new String[]{contactId} ,null);
                StringBuilder tel=new StringBuilder();
                //遍历查询结果，获取该联系人的所有电话号码
                while (numbers.moveToNext()) {
                    tel.append(numbers.getString(numbers.getColumnIndex(
                            ContactsContract.CommonDataKinds.Phone.NUMBER)));
                    if (!numbers.isLast()) tel.append("\n");
                }
                map.put("name", name);
                map.put("tel", tel);
                  //仅取联系人的第一个电话号码
                /* if (numbers.moveToNext()){
                    number=numbers.getString(numbers.getColumnIndex(
                            ContactsContract.CommonDataKinds.Phone.NUMBER));
                    map.put("tel", number);
                }*/
                contactsInfo.add(map);
            }
        }

    private void initView() {
        listView = (ListView) findViewById(R.id.list);
        queryContact();
        adapter = new SimpleAdapter(this, contactsInfo, R.layout.list, new String[]{"name", "tel"},
new int[]{R.id.tv_name,R.id.tv_tel});
        listView.setAdapter(adapter);
    }
    public void requestPermission() {
        int hasPermission = checkCallingOrSelfPermission(
                    Manifest.permission.READ_CONTACTS);
        if (hasWriteContactPermission != PackageManager.PERMISSION_GRANTED) {
            if (Build.VERSION.SDK_INT >= Build.VERSION_CODES.M) {
```

```
                    requestPermissions(new String[]{Manifest.permission.READ_CONTACTS}, 100);
                }
            }
        }
    }
```

程序运行结果如图 6.9 所示。

图 6.9 联系人信息

习题 6

1. 使用系统内置的 ContentProvider 获取手机中的短信。
2. 使用系统内置的 ContentProvider 获取手机相册中的图片。

第 7 章 服务

由于手机硬件性能和屏幕尺寸的限制，通常 Android 系统仅允许一个应用程序处于激活状态并显示在手机屏幕上，而暂停其他处于未激活状态的程序。因此，Android 系统需要一种后台服务机制，允许在没有用户界面的情况下使程序能够长时间在后台运行，并能够处理事件或更新数据。为此，Android 系统提供了 Service（服务）组件，专门用于执行持续性的、耗时且无须与用户界面交互的操作，如从网络下载文件、播放音乐、执行文件 I/O、与内容提供程序交互等。本章介绍 Service 组件及其生命周期，Service 的创建、启动，Service 与 Activity 之间的通信，系统服务的使用及异步消息处理等内容。

7.1 Service 简介

Service 与 Activity 是两个相似的组件，它们都代表可执行的程序。二者的区别是 Activity 可以显示用户界面并与用户交互，而 Service 没有用户界面。此外，Service 比 Activity 具有更高的优先级，在系统资源紧张的情况下 Service 被终止的可能性更小一些。即使被系统终止了，在系统资源恢复后，Service 也可以自动恢复运行状态。

按照运行的进程不同，可以将 Service 分为本地（Local）Service 和远程（Remote）Service。

本地 Service 运行于其客户端的应用程序进程中，当客户端终止（注意是客户端进程结束而不是 Activity 结束）后，本地 Service 也会终止。

在默认情况下，本地 Service 运行在其客户端应用程序进程的主线程中，也就是说，每当 Service 被调用，都会阻塞当前的 Activity 主线程，直到 Service 方法执行完毕，Activity 才有机会继续执行。因此，如果需要在 Service 中处理一些网络连接等耗时操作，应该将这些任务放在子线程中处理，以确保主线程能够保持畅通。

在实际应用中，有时希望 Service 作为后台服务，不仅被同一进程内的 Activity 使用，也可以被其他进程所使用，这就需要用到远程 Service。远程 Service 运行于独立的进程中，与其客户端之间需要进行跨进程通信，当客户端终止后，远程 Service 会继续进行。远程 Service 一般适用于为其他应用程序提供公共服务的 Service，这种 Service 即为系统常驻的 Service。

Android 系统提供了大量可以直接调用的系统 Service，如闹钟、振动、通知栏消息等，通过向这些 Service 传递特定的数据，可以方便地运行系统服务。

7.2 Service 的生命周期

与 Activity 类似，Service 也有生命周期，其生命周期与启动方式有关。Service 的启动方式有两种：一种是通过 startService()方法启动；另一种是通过 bingService()方法启动。使用不同的方式启动服务，执行的生命周期方法也不同。Service 的生命周期方法示意如图 7.1 所示。

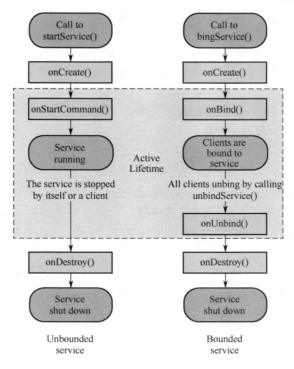

图 7.1　Service 的生命周期方法示意

从图 7.1 可以看出，当通过 startService()方法启动服务时，执行的生命周期方法依次为 onCreate()、onStartCommand()、onDestroy()。当通过 bingService()方法启动服务时，执行的生命周期方法依次为 onCreate()、onBind()、onUnbind()、onDestroy()。

无论使用哪种方式启动，Service 都会经历 onCreate()方法和 onDestroy()方法。通过 startService()方法启动服务时会调用 onStartCommand()方法，通过 bingService()方法启动服务时则会调用 onBind()方法，取消绑定时会调用 onUnbind()方法。

下面针对生命周期中的相关回调方法进行介绍。

1. onCreate()方法

创建服务时，系统将调用此方法。startService()方法启动服务，在服务未被创建时，系统会先调用 onCreate()方法创建服务。如果调用 startService()方法前服务已被创建，多次调用 startService()方法并不会多次创建服务。一般会在 onCreate()方法中完成必要的初始化工作，如创建线程、建立数据库等。

2. onStartCommand()方法

当一个组件（如 Activity）通过调用 startService()方法启动服务时，系统将调用 onStartCommand()方法。该 Service 将会一直在后台运行，而不管对应程序的 Activity 是否在运行，直到调用 stopService()方法，或者调用 Service 自身的 stopSelf()方法，或者启动程序被毁灭。当然如果系统资源不足，Android 系统也可能结束服务。如果一个 Service 被 startService() 方法多次启动时，onStartCommand() 方法将会被调用多次（对应调用 startService 的次数），但系统只会创建 Service 的一个实例。

onStartCommand() 方法的格式如下：

```
public int onStartCommand(Intent intent, int flags, int startId)
```

该方法的返回值决定当 Service 在资源不足的情况下被系统强行终止后的处理方式，其值可以是以下三种：

（1）START_NOT_STICKY。当 Service 因资源不足而被系统异常终止后，系统将不再自动重启 Service，除非再次调用 startService() 方法。这是最安全的选项，用来避免在不需要的时候运行服务。

（2）START_STICKY。当 Service 因资源不足而被系统异常终止后，在资源可用时系统将会尝试重新创建该 Service，由于服务状态保留在已开始状态，所以创建服务后一定会调用 onStartCommand() 方法。如果在此期间没有任何启动命令被传递到 Service，那么参数 Intent 将为 null。适用于不执行命令的媒体播放器（或类似的服务），它只是无限期的运行着并等待工作的到来。

（3）START_REDELIVER_INTENT。当 Service 因资源不足而被系统异常终止后，系统在资源可用时会重新创建此 Service 并调用 onStartCommand() 方法，并将 Service 被终止前收到的最后一个 Intent 对象传入 onStartCommand() 方法。适用于那些应该立即恢复正在执行的服务，如下载文件。

第一个参数 intent 为启动 Service 时所传入的 Intent 对象。

第二个参数 flags 的值代表 Service 的启动方式，与 onStartCommand() 方法的返回值有一定的关系。

正常启动 Service 时，将参数 flags 传入 0。

由上述 onStartCommand() 方法的返回值可知，在 Service 因资源不足被异常终止后，如果该 Service 的 onStartCommand() 方法返回值为 START_STICKY 或 START_REDELIVER_INTENT 时，系统会在资源可用时自动重新启动该 Service。

如果上次 onStartCommand() 方法的返回值为 START_STICKY，那么重新创建 Service，调用 onStartCommand() 方法时，将参数 intent 传入 null，参数 flags 传入 START_FLAG_RETRY。

如果上次 onStartCommand() 方法的返回值为 START_REDELIVER_INTENT，那么重新创建 Service，调用 onStartCommand() 方法时，将参数 flags 传入 START_FLAG_REDELIVERY。

第三个参数 startId 为启动请求 ID，用于唯一标识一次启动请求。每次调用 startService() 方法启动服务，onStartCommand() 方法都会被调用，就会产生一个请求 ID。在调用 stopSelfResult(int startId) 方法停止服务时可传入特定的 startId，可以更安全地根据 ID 停止服务。

采用 startService() 方法启动的服务，在服务完成后，可以通过服务自身调用 stopSelf() 或其他组件调用 stopService() 方法来停止服务。

3．onBind() 方法

当一个组件通过调用 bindService() 方法启动 Service 时，系统将调用此方法。

onBind()方法的格式如下：
```
public IBinder onBind(Intent intent)
```
参数 intent 为绑定这个 Service 时传入的 Intent 对象。返回值是一个 IBinder 对象，此对象定义了客户端与服务进行交互时所需的编程接口。在此方法的实现中，必须返回一个 IBinder 接口的实现类，供客户端与服务进行通信。

onBind()方法是 Service 类中的唯一的抽象方法，子类必须重写该方法。无论哪种方法启动服务，此方法必须重写，但在 startService()方法启动服务的情况下直接返回 null。

4．onUnbind()方法

断开服务绑定时调用此方法。当调用 unbindService()方法结束服务时，系统就会调用服务的 onUnbind()方法解除绑定。

5．onDestroy()方法

当服务不再使用且将被销毁时，系统会调用此方法。当调用 stopService()方法结束服务时系统会调用 onDestroy()方法；当调用 unbindService()方法结束服务时，系统会先调用 onUnbind()方法解除绑定，然后调用 onDestroy()方法销毁服务。一般在 onDestroy()方法中释放 Service 所占用的资源，如线程、注册的侦听器、接收器等。

7.3　Service 的使用

使用 Service 需要先创建一个子类继承 Service 类，并且重写 Service 类中的一些方法，然后在应用程序的 AndroidManifest.xml 文件中注册 Service 组件，并启动 Service。

7.3.1　创建和配置 Service

Service 与 Activity 很相似，有自己的生命周期，其创建、配置的方式也与 Activity 相似。首先在 Android Studio 中创建一个应用程序，然后在该应用程序的包名处单击右键，执行"New"→"Service"→"Service"菜单命令，在弹出的窗口中输入 Service 的名称 MyService，默认勾选"Exported"和"Enabled"复选框，单击"Finsh"按钮完成创建，如图 7.2 所示。

创建好的 Service 代码如下：
```
public class MyService extends Service {
    public MyService() {
    }
    @Override
    public IBinder onBind(Intent intent) {
        // TODO: Return the communication channel to the service.
        throw new UnsupportedOperationException("Not yet implemented");
    }
}
```

上述代码中，创建的 MyService 继承自 Service，默认一个构造方法 MyService()，并重写 onBind()方法。onBind()方法是 Service 子类必须实现的方法，该方法返回一个 IBinder 对象，应用程序可通过该对象与 Service 组件通信。

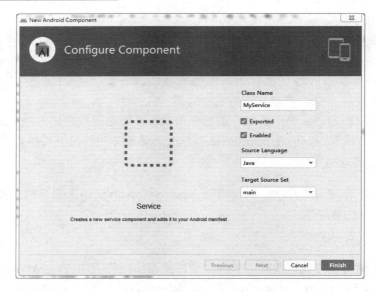

图 7.2 创建 Service

编写完成 Service 类后，还需要在应用程序的 AndroidManifest.xml 文件中配置 Service 组件，Android 才能在 APK 应用安装时解析出 Service 组件的信息，从而允许其他组件启动这个 Service。

在 AndroidManifest.xml 文件中，每个 Service 组件都需要在<application>标签的<service>子标签中进行配置。使用 Android Studio 向导创建 Service，AndroidManifest.xml 文件中自动添加了如下注册代码：

```
<service
    android:name=".MyService"
    android:enabled="true"
    android:exported="true">
</service>
```

上述代码中，<service>标签有三个属性，分别是 name、enabled、exported，其中，name 属性表示 Service 的名称，一般与创建的 Service 类名一致；enabled 属性表示系统是否能够实例化该 Service；exported 属性表示该 Service 是否能够被其他应用程序中的组件调用或进行交互。如果创建 Service 时没有勾选"Exported"和"Enabled"复选框，则这个属性的值为 false。除以上三个属性外，在 AndroidManifest.xml 文件中 Service 标签还有以下常见属性：

permission 声明此服务的权限，意味着只有提供了该权限的应用才能控制或连接此服务。

process 表示该服务是否运行在另外一个进程中。一般情况下 Service 没有自己独立的进程，仅运行于它所在的应用进程中。如果需要让一个 Service 在其他进程中运行，可以在通过 android:process 属性为其指定一个进程，如 android:process=":remote"。

若采用创建 Java 类继承 Service 类的方式创建服务，则需要在 AndroidManifest.xml 文件中手动添加代码进行注册，否则 Service 无法启动。

7.3.2 使用 startService()方法启动 Service

组件（如 Activity）通过调用 startService()方法启动 Service，称为启动方式。以这种方式启动的 Service 可在后台无限期运行,除非 Service 调用自身的 stopSelf()方法或其他组件调用 stopService()方法停止服务。当然在系统资源不足的情况下,Android 也会结束 Service。一个组件通过 startService()方法启动 Service 后,该组件和 Service 之间并没有关联,即使组件被销毁,也不影响 Service 的运行。

使用 startService()方法启动 Service,仅需要传递一个 Intent 对象即可,在 Intent 对象中指定需要启动的 Service。

```
Intent intent = new (MainActivity.this, MyService.class);
startService(intent);
```

使用 startService()方法启动服务,在服务的外部,使用 stopService()方法停止服务,在服务的内部可以调用自身 stopSelf()方法停止当前服务。使用 stopService()方法或 stopSelf()方法请求停止服务,系统就会销毁该服务。

【例 7.1】一个简单的后台音乐服务。

本例通过一个简单的后台音乐服务程序演示 Service 的创建、启动、销毁过程。

(1)建立 raw 文件夹。

首先创建一个应用程序,然后右击应用程序的 res 目录,在弹出的菜单中执行"New"→"Android Resource Directory"菜单命令,弹出"New Resource Directory"对话框,在"Resource Type"列表中选择"raw"选项,然后单击"OK"按钮,创建 raw 文件夹。

(2)准备音乐文件。

准备一个 MP3 文件,并将该音乐文件复制到应用程序的资源目录 res\raw 下。

(3)设计主界面。

在主界面上添加一个 TextView 控件和三个 Button 控件。主界面布局文件 activity_main.xml 代码如下:

```xml
<?xml version="1.0" encoding="utf-8"?>
<LinearLayout xmlns:android="http://schemas.android.com/apk/res/android"
    xmlns:tools="http://schemas.android.com/tools"
    android:layout_width="match_parent"
    android:layout_height="match_parent"
    android:orientation="vertical"
    android:padding="10dp"
    tools:context=".MainActivity">
    <TextView
        android:id="@+id/text"
        android:layout_width="match_parent"
        android:layout_height="wrap_content"
        android:textSize="30dp"
        android:text="" />
    <Button
        android:id="@+id/btn_start"
        android:layout_width="wrap_content"
        android:layout_height="wrap_content"
        android:textSize="30dp"
        android:text="启动后台音乐服务" />
    <Button
        android:id="@+id/btn_stop"
```

```xml
            android:layout_width="wrap_content"
            android:layout_height="wrap_content"
            android:textSize="30dp"
            android:text="关闭后台音乐服务" />
        <Button
            android:id="@+id/btn_close"
            android:layout_width="wrap_content"
            android:layout_height="wrap_content"
            android:textSize="30dp"
            android:text="关闭应用程序" />
</LinearLayout>
```

（4）编写后台服务程序。

在 MusicService.java 中编写服务程序。

```java
public class MusicService extends Service {
    MediaPlayer mediaPlayer;
    @Override
    //Service 中的抽象方法，子类必须要实现
    public IBinder onBind(Intent intent) {
        Log.i("msg","onBind()");
        return null;
    }
    @Override
    public boolean onUnbind(Intent intent) {
        Log.i("msg","onUnbind()");
        return super.onUnbind(intent);
    }
    @Override
    public void onCreate() {
        super.onCreate();
        mediaPlayer=MediaPlayer.create(getApplicationContext(), R.raw.happynewyear);
        Log.i("msg","onCreate()");
    }
    @Override
    public int onStartCommand(Intent intent, int flags, int startId) {
        Log.i("msg","onStartCommand()"+",flags:"+flags+",startId:"+startId);
        mediaPlayer.start();
        mediaPlayer.setLooping(true);    //循环播放音乐
        return super.onStartCommand(intent, flags, startId);
    }
    @Override
    public void onDestroy() {
        super.onDestroy();
        Log.i("msg","onDestroy()");
        if (mediaPlayer!=null){
            mediaPlayer.stop();
            mediaPlayer.release();
        }
    }
}
```

（5）编写 MainActivity。

```java
public class MainActivity extends AppCompatActivity implements View.OnClickListener{
    private Button start,stop,exit;
    private TextView textView;
    Intent intent;
    @Override
    protected void onCreate(Bundle savedInstanceState) {
        super.onCreate(savedInstanceState);
        setContentView(R.layout.activity_main);
        start=findViewById(R.id.btn_start);
        stop=findViewById(R.id.btn_stop);
        textView=findViewById(R.id.text);
        exit=findViewById(R.id.btn_close);
        start.setOnClickListener(this);
        stop.setOnClickListener(this);
        exit.setOnClickListener(this);
    }
    @Override
    public void onClick(View v) {
        switch (v.getId()){
            case R.id.btn_start:
                textView.setText(" start service ...");
                intent=new Intent(this,MusicService.class);
                startService(intent);
                break;
            case R.id.btn_stop:
                stopService(intent);
                break;
            case R.id.btn_close:
                finish();
                break;
        }
    }
}
```

（6）配置服务。

MusicService 配置代码：

```xml
<service
    android:name=".MusicService"
    android:enabled="true"
    android:exported="true">
</service>
```

运行应用程序，主界面如图 7.3 所示。

按如下步骤进行操作，注意音乐的播放情况并观察 LogCat 中的信息。

（1）单击图 7.3 中的"启动后台音乐服务"按钮，启动音乐服务，播放音乐，同时 LogCat 窗口输出如图 7.4 所示的日志信息。

第一次单击"启动后台音乐服务"按钮启动服务时，系统首先调用 onCreate()方法创建服务，然后调用 onStartCommand()方法，给参数 flags 传入的值为 0，startId 的值为 1。

（2）连续单击三次"启动后台音乐服务"按钮，LogCat 窗口的日志信息如图 7.5 所示。

图 7.3　主界面 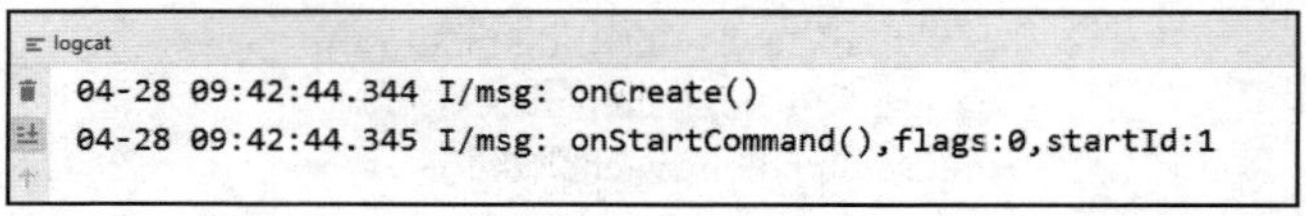　　　　　　　　　图 7.4　第一次启动 Service

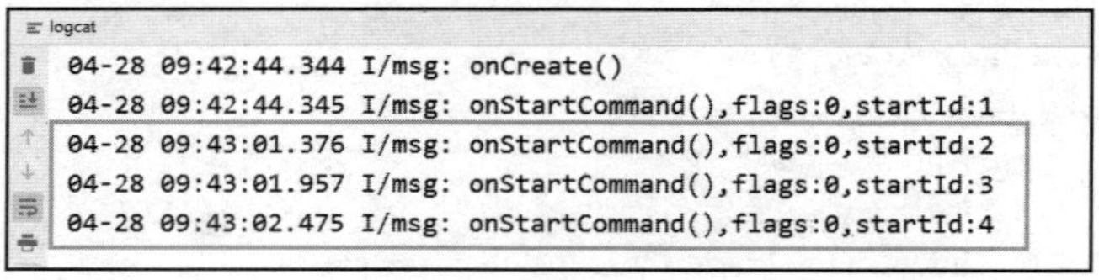

图 7.5　多次启动 Service

多次启动服务时，onStartCommand()方法会被调用多次，每次调用 onStartCommand()方法都会产生一个新的请求 ID，但系统只创建 Service 的一个实例。

（3）单击"关闭应用程序"按钮，应用程序退出，但是后台音乐仍然在播放。

（4）在模拟器首页找到应用程序，重新打开应用程序并单击"启动后台音乐服务"按钮，LogCat 窗口的日志信息如图 7.6 所示。

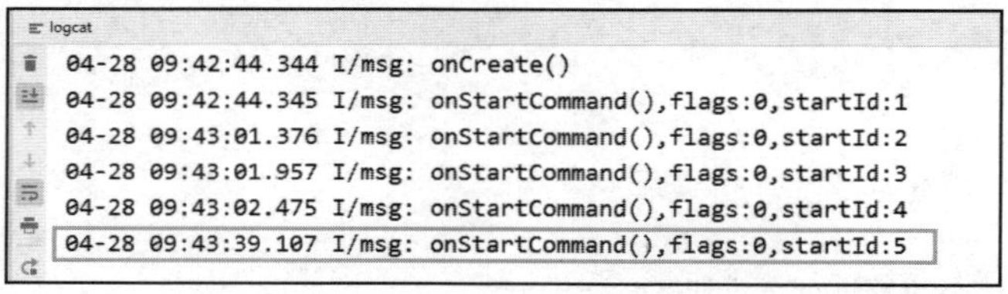

图 7.6　重新打开应用程序

由于服务已经被创建，所以重新打开应用程序并再次启动服务时，系统仅调用 onStartCommand()方法，产生一个新的请求 ID。

（5）单击主界面上的"关闭后台音乐服务"按钮，音乐停止播放，LogCat 窗口的日志信息如图 7.7 所示。

在"关闭后台音乐服务"按钮的单击事件中调用了 stopService(intent)方法停止服务，系统调用 onDestroy()方法销毁 Service 实例。

（6）单击"启动后台音乐服务"按钮再次启动服务，音乐开始播放，同时 LogCat 窗口的日志信息如图 7.8 所示。

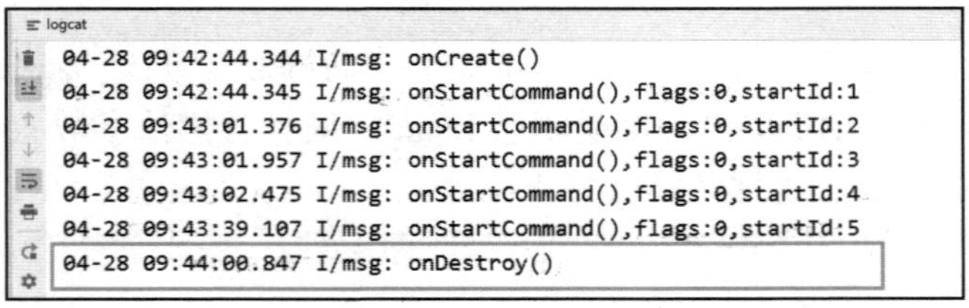

```
logcat
04-28 09:42:44.344 I/msg: onCreate()
04-28 09:42:44.345 I/msg: onStartCommand(),flags:0,startId:1
04-28 09:43:01.376 I/msg: onStartCommand(),flags:0,startId:2
04-28 09:43:01.957 I/msg: onStartCommand(),flags:0,startId:3
04-28 09:43:02.475 I/msg: onStartCommand(),flags:0,startId:4
04-28 09:43:39.107 I/msg: onStartCommand(),flags:0,startId:5
04-28 09:44:00.847 I/msg: onDestroy()
```

图 7.7　关闭后台音乐服务

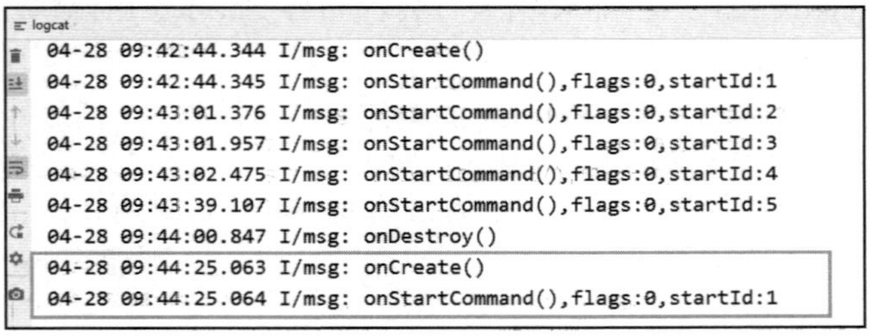

```
logcat
04-28 09:42:44.344 I/msg: onCreate()
04-28 09:42:44.345 I/msg: onStartCommand(),flags:0,startId:1
04-28 09:43:01.376 I/msg: onStartCommand(),flags:0,startId:2
04-28 09:43:01.957 I/msg: onStartCommand(),flags:0,startId:3
04-28 09:43:02.475 I/msg: onStartCommand(),flags:0,startId:4
04-28 09:43:39.107 I/msg: onStartCommand(),flags:0,startId:5
04-28 09:44:00.847 I/msg: onDestroy()
04-28 09:44:25.063 I/msg: onCreate()
04-28 09:44:25.064 I/msg: onStartCommand(),flags:0,startId:1
```

图 7.8　再次启动后台音乐服务

关闭音乐服务后再次启动服务，系统会调用 onCreate()方法重新创建服务并调用 onStartCommand()方法。

7.3.3　使用 bindService()方法启动 Service

使用 startService()方法启动的 Service 与启动其组件之间无任何关联，启动 Service 的组件不能获取 Service 的对象，因此无法调用 Service 的方法获取 Service 中的任何状态和数据信息。有时可能需要从 Activity 组件中去调用 Service 的方法，可以通过调用 bindService()方法启动 Service，这种启动方式又称为绑定方式。以这种方式启动的 Service 会与启动它的组件绑定，组件（如 Activity）可以向 Service 发送请求，或者调用 Service 的方法，被绑定的 Service 会接收信息并响应，可以通过绑定服务进行进程间通信。而且同一个 Service 可以绑定多个服务链接，这样可以同时为多个不同的组件提供服务。

绑定服务的 bindService()方法，其语法格式如下：

```
public boolean bindService(Intent service, ServiceConnection conn, int flags)
```

参数 service 表示绑定服务时传入的 Intent 对象，用于指定要启动的 Service。

参数 flags 表示绑定服务时是否自动创建 Service，该参数可设置为 0，表示不自动创建 Service，也可设置为 BIND_AUTO_CREATE，表示自动创建 Service。

参数 conn 是 ServiceConnection 接口类型的对象，用于监听调用者（服务绑定的组件）与 Service 之间的连接状态。当绑定成功时，系统将调用 ServiceConnection 的 onServiceConnected()方法。当绑定意外断开时，系统将调用 ServiceConnection 的 onServiceDisconnected()方法。

因此，以绑定方式使用 Service，调用者需提供一个 ServiceConnection 接口的实现类，并重写内部的 onServiceConnected()方法和 onServiceDisconnected()方法。

onServiceConnected()方法的格式如下：

```
public void onServiceConnected(ComponentName name, IBinder service)
```

其中，参数 name 为绑定 Service 的名称，参数 service 为绑定 Service 的 onBind()方法的

返回值。

Android 系统在创建调用者（如 Activity）与 Service 之间连接时，将调用 onServiceConnected()方法，并将 onBind()方法的返回值 IBinder 对象传递给参数 service。

onServiceDisconnected()方法的格式如下：

public void onServiceDisconnected(ComponentName name)

其中，参数 name 为绑定 Service 的名称。

不同于 startService()方法启动的服务默认无限期执行，bindService()方法启动的服务生命周期与其绑定的组件息息相关。当组件销毁时会自动与 Service 解除绑定，当没有任何组件与 Service 绑定时，Service 会自行销毁。

unbindService()方法用于解除与 Service 的绑定，其语法格式如下：

public void unbindService(ServiceConnection conn)

其中，参数 conn 是调用 bindService()方法绑定 Service 时所传入的 ServiceConnection 对象。需要注意的是，如果尚未绑定 Service 或已解除绑定，则调用 bindService()方法会抛出异常。

服务的两种启动方式并不是完全独立的，在某种情况下可以混合使用。以 MP3 播放为例，在后台工作的 Service 通过调用 startService()方法来启动，使它一直保持运行。但在播放过程中如果用户需要暂停音乐播放，则需要通过调用 bindService()方法获取服务链接和 Service 对象实例，进而通过调用 Service 对象实例中的方法来控制音乐播放。

7.3.4 Service 与 Activity 的通信

在 Android 中，Activity 负责前台页面的展示，Service 负责需要长期运行的任务，所以在实际开发中，常常遇到 Activity 与 Service 之间的通信。

在 Android 系统中，服务的通信方式有两种，即本地 Service 通信和远程 Service 通信。本地 Service 通信是指应用程序内部的通信，远程 Service 通信是指两个应用程序之间的通信。两种通信方式均需 Service 以绑定方式启动，否则无法进行通信和数据交换。

Android 中进程间通信依赖于 Android 接口定义语言 AIDL（Android Interface Definition Language）。AIDL 用于生成可以在 Android 设备上两个进程之间进行进程间通信（InterProcess Communication，IPC）的代码。

本地 Service 通信通过 IBinder 对象实现，其原理如图 7.9 所示。

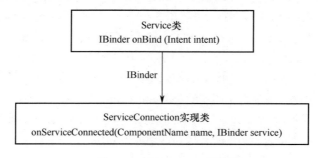

图 7.9 本地 Service 通信

当组件（客户端，如 Activity）调用 bindService()绑定一个 Service 时，Android 系统会调

用 Service 的 onBind()方法,该方法返回一个 IBinder 对象,Service 允许客户端通过该对象访问 Service 内部的数据,但这个 IBinder 对象不会直接返回给客户端,而是作为参数传递给 ServiceConnection 接口中的 onServiceConnected()方法,当服务绑定成功时系统回调该方法,在该方法中接收 Service 的 onBind()方法返回的 IBinder 对象。因此,要接收 IBinder 对象,客户端必须创建一个 ServiceConnection 接口实现类的对象并传给 bindService()方法。

实现本地 Service 通信还必须提供一个 IBinder 接口的实现类,然后在 onBind()方法中返回该类的一个实例,在 ServiceConnection 实现类中得到这个实例,用于客户端与服务器进行交互的编程接口。

IBinder 是用于进程内部和进程间过程调用的轻量级接口,定义了与远程对象交互的抽象协议,使用时可通过继承 Binder 的方法来实现。如果服务仅供本地应用使用,则不需要跨进程工作,可通过扩展 Binder 类来创建接口,并从 onBind()返回一个它的实例。客户端接收该 Binder 对象并用它来直接访问 Binder,以及 Service 中可用的公共(public)方法。

【例 7.2】一个简单的音乐播放器。

本例以一个简单的音乐播放器演示服务的创建、绑定、销毁,以及 Service 与 Activity 之间通信的过程。【例 7.1】使用 startService()方法启动音乐服务,音乐会在后台一直播放,直到播放完毕或停止服务,用户无法进行音乐的播放控制。本例使用 bindService()方法启动音乐服务,并且通过调用 Service 实例中的方法来控制音乐播放。

(1) 建立 raw 文件夹。

创建一个应用程序,并在 res 目录下建立 raw 文件夹。

(2) 准备音乐文件。

准备一个 MP3 文件,并将该音乐文件复制到应用程序的资源目录 res\raw 下。

(3) 设计主界面。

主界面布局文件 activity_main.xml 代码如下:

```xml
<?xml version="1.0" encoding="utf-8"?>
<LinearLayout xmlns:android="http://schemas.android.com/apk/res/android"
    xmlns:app="http://schemas.android.com/apk/res-auto"
    xmlns:tools="http://schemas.android.com/tools"
    android:layout_width="match_parent"
    android:layout_height="match_parent"
    android:padding="10dp"
    android:orientation="vertical"
    tools:context=".MainActivity">
    <LinearLayout
        android:layout_marginTop="50dp"
        android:layout_width="match_parent"
        android:layout_height="wrap_content"
        android:orientation="horizontal">
        <Button
            android:id="@+id/btn_play"
            android:layout_width="wrap_content"
            android:layout_height="wrap_content"
            android:textSize="20sp"
            android:layout_weight="1"
            android:text="播放" />
```

```xml
        <Button
            android:id="@+id/btn_pause"
            android:layout_width="wrap_content"
            android:layout_height="wrap_content"
            android:textSize="20sp"
            android:layout_weight="1"
            android:enabled="false"
            android:text="暂停"   />
        <Button
            android:id="@+id/btn_stop"
            android:layout_width="wrap_content"
            android:layout_height="wrap_content"
            android:textSize="20sp"
            android:layout_weight="1"
            android:enabled="false"
            android:text="停止"   />
    </LinearLayout>
    <Button
        android:id="@+id/btn_quit"
        android:layout_width="match_parent"
        android:layout_height="wrap_content"
        android:textSize="20sp"
        android:text="关闭音乐播放器"
         />
</LinearLayout>
```

（4）创建 MusicService 类。

在 MusicService 的生命周期方法中输出日志信息，定义内部类 MusicBinder，其中包含一个 getService()方法用于获取 MusicService 类的实例，在 onBind()方法中返回 MusicBinder 类的实例。该类中还包括播放控制方法 play()、pause()和 stop()。

MusicService.java 代码如下：

```java
public class MusicService extends Service {
    MediaPlayer player;
        @Override
        public void onCreate() {    //初始化 player
            super.onCreate();
            Log.i("msg", "onCreate()");
            player= MediaPlayer.create(this,R.raw.happynewyear);
        }
        @Override
        public void onDestroy() {    //销毁 player
            super.onDestroy();
            Log.i("msg ", "onDestroy()");
            if (player!=null){
                player.stop();
                player.release();    //释放 MediaPlayer 对象
            }
        }
        @Override
        public boolean onUnbind(Intent intent) {
```

```java
            Log.i("msg ", "onUnbind()");
            return super.onUnbind(intent);
        }
        @Override
        public IBinder onBind(Intent intent) {
            Log.i("msg ", "onBind()");
            return new MusicBinder();
        }
        class MusicBinder extends Binder {
            public MusicService getService() {
                    return MusicService.this;
            }
        }
        public void play(){
            if (player==null){
                player=MediaPlayer.create(this,R.raw.happynewyear);
                player.start();
                return;
            }
            if (player!=null && !player.isPlaying()){    //暂停状态
                player.start();
                return;
            }
             if (player!=null && player.isPlaying()){    //播放状态, 重新播放
                player.seekTo(0);
                player.start();
                return;
            }
        }
        public void pause(){
            if (player!=null && player.isPlaying()){
                player.pause();
            }
        }
        public void stop(){
            if (player!=null && player.isPlaying()){
                player.stop();
                player.prepareAsync();
            }
        }
    }
```

（5）编写 MainActivity 中的代码。

在 MainActivity 中定义内部类 MyConn，重写 onServiceConnected()方法，onBind()方法返回的 MusicBinder 类的实例通过 onServiceConnected()方法的第二个参数接收，并通过它获取 MusicService 实例来访问 Service 中播放控制方法。在 MainActivity 的 onCreate()方法中调用 bindService()方法启动服务，并在 onDestroy()方法中解除与 Service 的绑定。

```java
public class MainActivity extends AppCompatActivity implements View.OnClickListener{
    private Button btn_play,btn_pause,btn_stop,btn_quit;
    MyConn conn=new MyConn();
```

```java
        MusicService musicService;
            @Override
            protected void onCreate(Bundle savedInstanceState) {
                super.onCreate(savedInstanceState);
                setContentView(R.layout.activity_main);
                initView();
                Intent intent =new Intent(this,MusicService.class);
                bindService(intent, conn, BIND_AUTO_CREATE);
            }
            public void initView() {
                btn_play=findViewById(R.id.btn_play);
                btn_pause=findViewById(R.id.btn_pause);
                btn_stop=findViewById(R.id.btn_stop);
                btn_quit=findViewById(R.id.btn_quit);
                btn_play.setOnClickListener(this);
                btn_pause.setOnClickListener(this);
                btn_stop.setOnClickListener(this);
                btn_quit.setOnClickListener(this);
            }
            @Override
            public void onClick(View v) {
                switch (v.getId()){
                    case R.id.btn_play:
                        musicService.play();
                        btn_play.setEnabled(false);
                        btn_pause.setEnabled(true);
                        btn_stop.setEnabled(true);
                        break;
                    case R.id.btn_pause:
                        musicService.pause();
                        btn_play.setEnabled(true);
                        btn_pause.setEnabled(false);
                        btn_stop.setEnabled(true);
                        break;
                    case R.id.btn_stop:
                        musicService.stop();
                        btn_play.setEnabled(true);
                        btn_pause.setEnabled(false);
                        btn_stop.setEnabled(false);
                        break;
                    case R.id.btn_quit:
                        finish();
                        break;
                }
            }
            @Override
            protected void onDestroy() {
                super.onDestroy();
                unbindService(conn);
            }
            class MyConn implements ServiceConnection{
```

```
            @Override
            public void onServiceConnected(ComponentName name, IBinder service) {
                //获取 MusicBinder 的实例
        MusicService.MusicBinder binder=(MusicService.MusicBinder) service;
                musicService=binder.getService();   //返回一个 MusicService 对象
            }
            @Override
            public void onServiceDisconnected(ComponentName name) {

            }
        }
```

在 onServiceConnected(ComponentName name, IBinder service)回调方法中,将第二个参数 service 强制转换成 MusicBinder 对象,通过其 getService()方法得到一个 MusicService 对象,然后通过它调用 MusicService 中的方法。

运行应用程序,初始主界面如图 7.10(a)所示,LogCat 窗口的日志信息如图 7.11 所示。

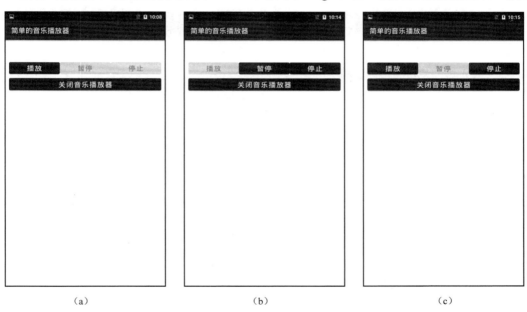

图 7.10　主界面

由于在 MainActivity 的 onCreate()方法中调用 bindService()方法启动服务,所以应用程序启动时系统依次调用 Service 的 onCreate()方法和 onBind()方法创建并绑定服务。

单击"播放"按钮,开始播放音乐,同时"暂停"按钮和"停止"按钮变为可操作,"播放"按钮变为不可操作,如图 7.10(b)所示。

单击"暂停"按钮,音乐暂停播放,同时"播放"按钮变为可操作,如图 7.10(c)所示。

再次单击"播放"按钮,继续播放音乐,各按钮状态如图 7.10(b)所示。

单击"停止"按钮,停止播放音乐,各按钮状态如图 7.10(a)所示。

单击"关闭音乐播放器"按钮,应用程序退出,音乐停止,同时 LogCat 窗口的日志信息如图 7.12 所示。

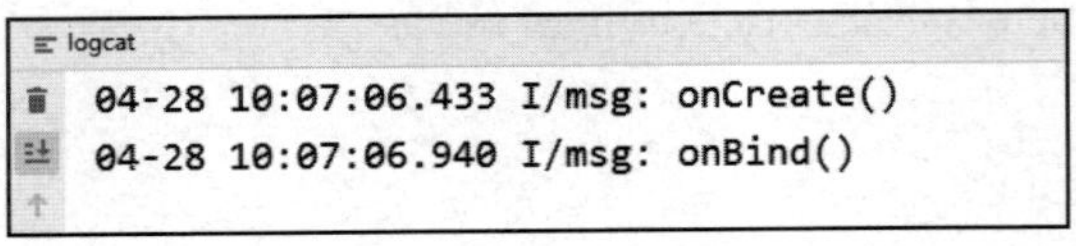

图 7.11　运行音乐播放器 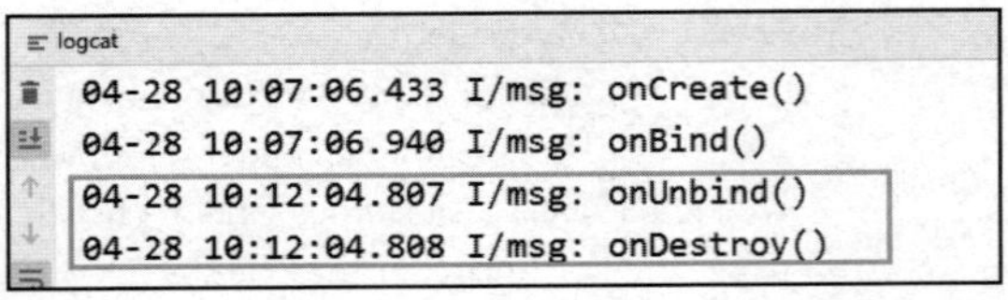　　　　　图 7.12　关闭音乐播放器

由于在 MainActivity 的 onDestroy()方法中调用了 unbindService()方法解除与 Service 的绑定，所以在应用程序退出时系统首先调用 Service 的 onUnbind()方法解除绑定，然后调用 onDestroy()方法销毁服务。

7.4　访问系统服务

Android 提供了许多系统级别的 Service，通过这些 Service，应用程序可以方便地调用系统功能。通过 Context.getSystemService（String serviceName）获取系统服务，其中参数 serviceName 表示需要的服务名称，而系统服务的名称都在 Context 类中定义了常量。

通过 getSystemService()方法会返回一个特定服务的管理器对象，使用此对象可以完成服务调用功能。常用的系统服务如表 7.1 所示。

表 7.1　常用的系统服务

服 务 对 象	服务名称常量	功　　能
AlarmManager	ALARM_SERVICE	闹钟服务
AudioManager	AUDIO_SERVICE	音频服务
BluetoothAdapter	BLUETOOTH_SERVICE	蓝牙服务
ConnectivityManager	CONNECTIVITY_SERVICE	网络连接服务
DownloadManager	DOWNLOAD_SERVICE	针对 HTTP 的下载服务
NotificationManager	NOTIFICATION_SERVICE	通知服务
LocationManager	LOCATION_SERVICE	GPS 定位服务等
PowerManager	POWER_SERVICE	电源服务
SensorManager	SENSOR_SERVICE	传感器服务
Vibrator	VIBRATOR_SERVICE	振动器服务
WifiManager	WIFI_SERVICE	Wi-Fi 服务

下面以常用的系统通知服务和手机闹钟服务为例介绍系统服务的使用方法。

1．NotificationManager（系统通知服务）

Android 系统提供的消息提示机制有 Toast 和 Notification。Toast 是一种快速的即时消息，消息的内容简短，悬浮于应用程序的最上方。Notification 是一种具有全局效果的通知，以图标的形式显示在系统的通知栏中，用户可以下拉通知栏查看通知的详细信息。Notification 还可以使用其他形式提示信息，如闪烁 LED、振动、发出声音等，也可启动另一个 Intent。

所有的 Notification 通过 NotificationManager 来管理。使用 Notification 的步骤如下。

（1）获取 NotificationManager 对象。

NotificationManager manager = (NotificationManager)

```
                    getSystemService(NOTIFICATION_SERVICE);
```
（2）创建通知。
```
Notification notification = new Notification.Builder(this)
    .setSmallIcon(R.drawable.ic_launcher_background)
    .setLargeIcon(BitmapFactory.decodeResource(getResources(), R.drawable.logo))
    .setContentTitle("紧急通知")
    .setContentText("沙尘暴预警")
    .build();
```
（3）发送系统通知。
```
manager.notify(notifyId, builder.build());
```
（4）取消通知。
```
manager.cancel(notifyId);    //取消 notifyId 关联的通知
manager.cancelAll();    //取消所有通知
```

【例 7.3】在通知栏中显示系统通知。

（1）界面设计。

在 activity_main.xml 中添加两个 Button 控件，代码如下：

```xml
<?xml version="1.0" encoding="utf-8"?>
<LinearLayout xmlns:android="http://schemas.android.com/apk/res/android"
    xmlns:tools="http://schemas.android.com/tools"
    android:layout_width="match_parent"
    android:layout_height="match_parent"
    android:padding="20sp"
    android:orientation="vertical"
    tools:context=".MainActivity">
    <Button
        android:id="@+id/btn_send"
        android:layout_width="wrap_content"
        android:layout_height="wrap_content"
        android:textSize="30dp"
        android:text="发送系统通知" />
    <Button
        android:id="@+id/btn_delete"
        android:layout_width="wrap_content"
        android:layout_height="wrap_content"
        android:textSize="30dp"
        android:text="删除通知" />
</LinearLayout>
```

（2）编写 MainActivity。

在 MainActivity 中定义一个 showNotification()方法，用于创建并发送通知。

```java
public class MainActivity extends AppCompatActivity implements View.OnClickListener {
    private Button btn_send,btn_delete;
    NotificationManager mNotifyMgr;
    Notification notification;
    @Override
    protected void onCreate(Bundle savedInstanceState) {
        super.onCreate(savedInstanceState);
        setContentView(R.layout.activity_main);
        btn_send=findViewById(R.id.btn_send);
        btn_delete=findViewById(R.id.btn_delete);
        btn_send.setOnClickListener(this);
```

```java
            btn_delete.setOnClickListener(this);
        }
        private void showNotification(){
            mNotifyMgr=(NotificationManager) getSystemService(NOTIFICATION_SERVICE);
            Intent resultIntent = new Intent(this, MainActivity.class);
            PendingIntent resultPendingIntent = PendingIntent.getActivity(
                    this, 0, resultIntent, PendingIntent.FLAG_UPDATE_CURRENT);
            notification = new Notification.Builder(this)
                    .setSmallIcon(R.drawable.ic_launcher_background)
                        .setLargeIcon(BitmapFactory.decodeResource(getResources(), R.drawable.logo))
                    .setContentTitle("紧急通知")
                    .setContentText("沙尘暴预警")
                    .setAutoCancel(true)    //单击删除通知
                    .setContentIntent(resultPendingIntent)
                    .build();
            mNotifyMgr.notify(0, notification);
        }
        @Override
        public void onClick(View v) {
            switch (v.getId()){
                case R.id.btn_send:
                    showNotification();
                    break;
                case R.id.btn_delete:
                    mNotifyMgr.cancelAll();
                    break;
            }
        }
    }
```

运行程序，主界面如图 7.13 所示。单击"发送系统通知"按钮，在系统通知栏中以图标形式显示通知，向下拖动通知栏可以查看通知信息，如图 7.14 所示，单击"删除通知"按钮即可。

图 7.13　主界面

图 7.14　查看通知

2. AlarmManager（手机闹钟服务）

AlarmManager 本质上是一个全局定时器，可以在指定时间或指定周期启动相应的组件，包括 Activity、Service 和 BroadcastReceiver。AlarmManager 可以通过 context. getSystem (ALARM_SERVICE)来获取，调用其包含的方法来设置定时启动的组件。

使用 set(int type, long triggerAtMills, PendingIntent operation)方法设定一次性闹钟，其中，type 参数是闹钟的唤醒方式，通常使用 RTC_WAKEUP；triggerAtMills 参数为闹钟执行时间，可以从 TimePicker 中获取，在 onTimeSet 接口中将用户设定的时间值保存在一个 Calendar 对象中，之后通过 getTimeInMills 获取定时时间赋值给 triggerAtMills；operation 参数为 PendingIntent 类型，代表闹钟服务执行的动作，通过该参数指定闹钟执行时要启动的组件，可以是一个 Activity，也可以是 Service 或 Broadcast。PendingIntent 是 Intent 的封装类，如在闹钟执行时唤醒 ClockActivity，代码如下：

```
Intent intent = new Intent(this, ClockActivity.class);
PendingIntent pendingIntent = PendingIntent.getActivity(this, 1, intent, PendingIntent.FLAG_UPDATE_CURRENT);
```

如果闹钟服务执行时要唤醒一个 Service，则用 getService()方法；如果发送一个广播，则用 getBroadcast()方法。

从 API 19 开始，系统为了减少唤醒和电量使用，对闹钟时间进行了一定的偏移，因此闹钟唤醒的时间是不准确的。Android 提供了新的 APIs 来支持那些需要准时唤醒的应用，可以参考以下两种方法：

```
setExact(int type, long triggerAtMills, PendingIntent operation)
setWindow(int type, long windowStartMills, long windowLengthMills, PendingIntent operation)
```

【例 7.4】使用 AlarmManager 定时启动音乐服务。

本例通过 AlarmManager 定时启动后台音乐服务 MusicService。

（1）创建 raw 文件夹。

创建一个应用程序，在应用程序的 res 目录下建立 raw 文件夹，并将音乐文件复制到 raw 中。

（2）创建 MusicService。

代码同【例 7.2】中的 MusicService。

（3）界面设计。

在 activity_main 布局中添加两个 Button 控件，activity_main.xml 代码如下：

```xml
<?xml version="1.0" encoding="utf-8"?>
<LinearLayout xmlns:android="http://schemas.android.com/apk/res/android"
    android:orientation="vertical"
    android:padding="20sp"
    xmlns:tools="http://schemas.android.com/tools"
    android:layout_width="match_parent"
    android:layout_height="match_parent"
    tools:context=".MainActivity">
    <Button
        android:id="@+id/btn_set"
        android:layout_width="match_parent"
        android:layout_height="wrap_content"
        android:textSize="30dp"
```

```xml
            android:text="设置闹钟" />
        <Button
            android:id="@+id/bn_cancel"
            android:layout_width="match_parent"
            android:layout_height="wrap_content"
            android:textSize="30dp"
            android:text="关闭闹钟" />
</LinearLayout>
```

（4）编写 MainActivity。

在 MainActivity 中定义监听器类 timeListener 用于 TimePickerDialog 控件的监听器，在"设置闹钟"按钮的单击事件中通过 TimePickerDialog 设定时间。

```java
public class MainActivity extends AppCompatActivity implements View.OnClickListener {
    private Button alarmTime,cancel;
    private AlarmManager alarmManager;
    private PendingIntent pendingIntent;
    @Override
    protected void onCreate(Bundle savedInstanceState) {
        super.onCreate(savedInstanceState);
        setContentView(R.layout.activity_main);
        alarmTime = findViewById(R.id.btn_set);
        cancel=findViewById(R.id.bn_cancel);
        cancel.setOnClickListener(this);
        alarmTime.setOnClickListener(this);
    }
    @Override
    public void onClick(View v) {
        switch (v.getId()){
            case R.id.btn_set:    //通过 TimePickerDialog 设定时间
                Calendar current=Calendar.getInstance();
                int hour=current.get(Calendar.HOUR_OF_DAY);
                int minute=current.get(Calendar.MINUTE);

                new TimePickerDialog(MainActivity.this,0,
                        new timeListener(),hour,minute,true).show();
                break;
            case R.id.bn_cancel:
                alarmManager.cancel(pendingIntent);    //取消闹钟
                break;
        }
    }
    //监听器类
    class timeListener implements TimePickerDialog.OnTimeSetListener {
        @Override
        public void onTimeSet(TimePicker view, int hourOfDay, int minute) {
            Calendar calendar=Calendar.getInstance();
            calendar.setTimeInMillis(System.currentTimeMillis());
            calendar.set(Calendar.HOUR, hourOfDay);
            calendar.set(Calendar.MINUTE, minute);
            Intent intent=new Intent(MainActivity.this,MusicService.class);
```

```
            pendingIntent=PendingIntent.getService(MainActivity.this, 0, intent, 0);
            alarmManager=(AlarmManager)getSystemService(Service.ALARM_SERVICE);
            alarmManager.set(AlarmManager.RTC_WAKEUP, calendar.getTimeInMillis(), pendingIntent);
        }
    }
}
```

运行程序，主界面如图 7.15 所示，单击"设置闹钟"按钮，弹出如图 7.16 所示的时间选择对话框，设置闹钟时间，定时时间到后台音乐就会开始播放。

图 7.15　主界面　　　　　　　　图 7.16　设置定时

7.5　异步消息处理

服务在其托管进程的主线程中运行（UI 线程），这意味着，如果服务执行耗时操作或可能被阻塞的任务（如 MP3 播放或联网），将会造成界面无响应甚至出现 ANR 错误。因此服务执行耗时操作时，应在服务内创建子线程来完成。在 Android 中，子线程不能更新 UI，子线程操作完成后需要将操作结果发送给主线程进行更新。所以，子线程和主线程是异步操作的，Android 提供了两种多线程异步操作的实现方式。

7.5.1　Handler 消息传递机制

当应用程序启动时，Android 会启动一条主线程（Main Thread），负责处理与 UI 相关的事件，如用户按钮事件，并把相关的事件分发到对应的组件进行处理后再绘制界面，因此主线程又称为 UI 线程。

但主线程不能做耗时操作，这是因为 Android 系统是单线程模型，系统给 App 分配的进程里，只有一个主线程，所有的组件都在主线程里实现、调度和管理，组件的绘制、互动操作和生命周期回调都是主线程处理的。如果这个线程出现阻塞、多线程死锁、CPU 饥饿都将导致事件处理时间过长，甚至没有机会得到处理，系统就会处理为 ANR，强制关闭 App，所以耗时操作需要开启子线程来处理。

然而，出于性能优化的考虑，当 Android UI 操作时，如果有多个线程并发操作 UI 组件，就可能导致线程安全问题。因此，Android 规定只允许主线程进行 UI 操作，不允许子线程更新 UI。当子线程完成任务时，需要告知主线程进行 UI 更新，这需要借助 Handler 消息传递机制来实现，子线程与主线程通过 Handler 来进行通信。

Handler 机制是 Android 用来处理异步消息的核心类。一般情况下，主线程中绑定了 Handler 对象，并在事件触发上创建子线程用于完成某些耗时操作。当子线程中的工作完成之后，向 Handler 对象发送一个已完成的信号（Message 对象），当 Handler 对象接收到消息后，就会对主线程 UI 进行更新操作。

Handler 机制主要包括四个关键对象，分别是 Message、Handler、MessageQueue 和 Looper。

1. Message 对象

Message 对象是在线程之间传递的消息，它可以携带少量的信息，用于在不同线程之间交换数据。Message 对象的主要属性如下。

what 属性：Int 类型，存放消息 ID（Send 方式），主线程用其识别子线程发来的消息类型。

arg1 属性和 arg2 属性：Int 类型，如果传递的消息类型为 Int 型，则可以将数字赋给 arg1 和 arg2。

obj 属性：Object 类型，Message 对象所携带的消息体。如果传递的消息是 String 或其他，则可以赋给 obj。

2. Handler 对象

Handler 对象是 Message 对象的主要处理者，用来发送和处理消息。子线程通过 Handler 对象发送 Message 对象，Handler 对象负责将 Message 对象添加到消息队列，以及对 Looper 交付的 Message 对象进行处理。

3. MessageQueue 对象

MessageQueue 对象用来存放 Handler 对象发送过来 Message 对象。子线程通过 Handler 对象发送的 Message 对象会存放在 MessageQueue 对象中等待处理。每个线程只有一个 MessageQueue 对象。

4. Looper 对象

Looper 对象负责对 MessageQueue 对象进行管理，不断从 MessageQueue 对象中取出 Message 对象交给 Handler 对象处理。主线程启动时，会自动创建一个 Looper 对象，Looper 对象的构造方法会自动创建一个 MessageQueue 对象，然后通过循环不断地从 MessageQueue 对象中读取 Message 对象，并分发给相应的 Handler 对象进行处理。

如果在子线程中使用 Handler 对象，必须自己创建一个 Looper 对象并启动它。

Handler 对象的消息机制的原理如图 7.17 所示。

Handler 对象调用 sendMessage(message)方法发送 Message 对象给 MessageQueue 对象，Looper 对象不断地从 MessageQueue 对象中取出 Message 对象并分发给相应的 Handler 对象，Handler 对象调用 handleMessage(Message msg)方法对消息进行处理。

在主线程中使用 Handler 对象传递消息，首先需要在主线程中创建一个 Handler 对象，然后在子线程中调用该 Handler 对象的 sendMessage(message)方法发送 Message 对象，这个 Message 对象会存放在主线程的 MessageQueue 对象中，最终经过 Looper 对象取出后分发到相应 Handler 对象的 handleMessage(Message msg)方法中，主线程通过重写 Handler 对象的 handleMessage(Message msg)方法来接收这个 Message 对象并进行处理。

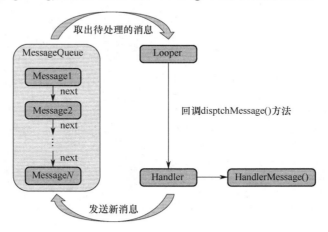

图 7.17　Handler 对象的消息机制

在主线程中使用 Handler 对象的步骤如下。

（1）在主线程中创建 Handler 对象。

```
public class MainActivity extends Activity {
    ...
    private Handler handler = new Handler() {
        @Override
        public void handleMessage(Message msg) {
            super.handleMessage(msg);
        }
    };
    ...
}
```

（2）在子线程中发送消息。

```
new Thread(new Runnable() {
    @Override
    public void run() {
        Message message = handler.obtainMessage();
        message.arg1 = 1;
        handler.sendMessage(message);
    }
}).start();
```

（3）在 Handler 对象中捕获消息。

```
public class MainActivity extends Activity {
    ...
    private Handler handler = new Handler() {
        @Override
```

```
            public void handleMessage(Message msg) {
                super.handleMessage(msg);
                if (msg.arg1 == 1) {
                    Toast.makeText(MainActivity.this, "hanlder",
                            Toast.LENGTH_SHORT).show();
                }
            }
        };
        ...
    }
```

【例 7.5】计时器。

本例通过一个计时器演示 Handler 对象消息机制的工作原理。

（1）界面设计。

```xml
<?xml version="1.0" encoding="utf-8"?>
<LinearLayout xmlns:android="http://schemas.android.com/apk/res/android"
    xmlns:tools="http://schemas.android.com/tools"
    android:layout_width="match_parent"
    android:layout_height="match_parent"
    android:orientation="vertical"
    tools:context=".MainActivity">
    <LinearLayout
        android:layout_marginTop="30dp"
        android:layout_width="match_parent"
        android:layout_height="wrap_content"
        android:padding="20sp"
        android:orientation="horizontal">
        <TextView
            android:id="@+id/textView2"
            android:layout_width="0dp"
            android:layout_height="wrap_content"
            android:gravity="center"
            android:layout_weight="3"
            android:textSize="30dp"
            android:text="输入计时时间" />
        <EditText
            android:id="@+id/edit_time"
            android:layout_width="0dp"
            android:layout_height="wrap_content"
            android:layout_weight="1"
            android:gravity="center"
            android:inputType="number"
            android:text="" />
        <TextView
            android:id="@+id/textView3"
            android:layout_width="0dp"
            android:layout_height="wrap_content"
            android:layout_weight="1"
            android:textSize="30dp"
            android:text="秒" />
    </LinearLayout>
    <Button
```

```xml
        android:id="@+id/button"
        android:layout_marginTop="30sp"
        android:layout_width="wrap_content"
        android:layout_gravity="center"
        android:textSize="30dp"
        android:layout_height="wrap_content"
        android:text="开始计时" />
    <TextView
        android:id="@+id/textView"
        android:padding="20sp"
        android:layout_width="match_parent"
        android:layout_height="wrap_content"
        android:textSize="30dp"
        android:layout_gravity="center"
        android:visibility="gone"
        android:text="计时时间：" />
</LinearLayout>
```

（2）编写 MainActivity。

在按钮的单击事件中开辟子线程，每秒钟发送一个消息。在 MainActivity 中创建一个 handler 对象，重写 handleMessage(Message msg)方法，处理子线程发来的消息，实现 UI 更新。

```java
public class MainActivity extends AppCompatActivity implements View.OnClickListener {
    private int time;
    private EditText editText;
    private TextView textView;
    private Button button;
//创建 handler 对象
    Handler handler=new Handler(){
        @Override
        public void handleMessage(Message msg) {    //处理消息
            super.handleMessage(msg);
            int arg1=msg.arg1;
            String info=(String) msg.obj;
            if (msg.what==1){
                textView.setText(info+arg1);
                if (arg1==time){
                    Toast.makeText(MainActivity.this,"计时结束！",
                            Toast.LENGTH_LONG).show();
                }
            }
        }
    };
    @Override
    protected void onCreate(Bundle savedInstanceState) {
        super.onCreate(savedInstanceState);
        setContentView(R.layout.activity_main);
        initView();
    }
    private void initView() {
        editText = (EditText) findViewById(R.id.edit_time);
        textView = (TextView) findViewById(R.id.textView);
        button = (Button) findViewById(R.id.button);
```

```
                button.setOnClickListener(this);
        }
        @Override
        public void onClick(View v) {
                textView.setVisibility(View.VISIBLE);
                time=Integer.valueOf(editText.getText().toString());
                textView.setText("计时: "+0);
                //创建子线程
                new Thread(){
                        @Override
                        public void run() {
                                for (int i=0;i<=time;i++){
                                        try{
                                                Thread.sleep(1000);
                                        }catch (Exception e){
                                                e.printStackTrace();
                                        }
                                        Message message=handler.obtainMessage();
                                        message.what=1;
                                        message.obj="计时: ";
                                        message.arg1=i;
                                        handler.sendMessage(message);
                                }
                        }
                }.start();
        }
}
```

运行程序，在如图 7.18（a）所示的界面输入计时时间，单击"开始计时"按钮，主界面会不断刷新计时时间，如图 7.18（b）所示，直到计时结束，如图 7.18（c）所示。

图 7.18　计时器

7.5.2　AsyncTask 类

Handler 机制为多线程异步消息处理提供了一种完善的处理方式,但是在处理较为简单的应用时,Handler 机制的使用稍显复杂,需要继承 Handler 类,重写处理方法,还需要开发者自己开辟线程,在线程中完成操作以后再发送消息将结果更新到 UI 线程。为了简化操作,Android 提供了一个轻量级异步类 AsyncTask,它可以在线程池中执行后台任务,然后把执行的进度和最终结果传递给主线程,并在主线程中更新 UI。

AsyncTask 类属于抽象类,使用时需实现子类。AsyncTask 的类声明:

```
public abstract class AsyncTask<Params, Progress, Result> {
    ...
}
```

AsyncTask 是一个泛型类,三个泛型类型参数的含义如下。
- Params:开始异步任务执行时传入的参数类型,对应 excute()方法中传递的参数。
- Progress:异步任务执行过程中,下载进度值的类型。
- Result:异步任务执行完成后,返回值的类型,与 doInBackground()的返回值类型一致。

使用时并不是所有类型都被使用,若没有被使用的,可用 java.lang.Void 类型代替。

使用 AsyncTask 可以重写以下四个回调函数。

(1) protected abstract Result doInBackground(Params... params)

用于执行后台任务,AsyncTask 的子类必须重写该方法。

(2) protected void onPreExecute()

用于执行一些初始化操作,在后台任务开始执行之前调用。

(3) protected void onPostExecute(Result result)

用于更新 UI 且显示结果,在 doInBackground 之后执行。

(4) protected void onProgressUpdate(Progress... values)

用于更新进度信息。

异步执行任务中系统会自动调用 AsyncTask 的一系列方法。

【例 7.6】模拟下载任务。

本例模拟一个下载任务,在 doInBackground()方法中执行具体的下载逻辑,先在 onProgressUpdate()方法中显示当前的下载进度,并在 onPostExecute()方法中提示任务的执行结果。

(1) 界面设计。

在主界面上,有一个 Button,一个 ProgressBar,一个 TextView。单击"Button"按钮,执行下载任务,下载过程中用 ProgressBar 和 TextView 显示下载的进度,下载完成后在 TextView 中显示"下载完成"。

Activity_main.xml 代码如下:

```xml
<?xml version="1.0" encoding="utf-8"?>
<LinearLayout xmlns:android="http://schemas.android.com/apk/res/android"
    xmlns:app="http://schemas.android.com/apk/res-auto"
    xmlns:tools="http://schemas.android.com/tools"
    android:layout_width="match_parent"
    android:layout_height="match_parent"
    android:orientation="vertical"
```

```xml
        tools:context=".MainActivity">
    <Button
        android:id="@+id/download"
        android:layout_width="300dp"
        android:layout_gravity="center"
        android:layout_marginTop="100dp"
        android:layout_height="wrap_content"
        android:text="Download" />
    <TextView
        android:id="@+id/tv"
        android:layout_width="100dp"
        android:layout_height="wrap_content"
        android:text="单击下载"
        android:layout_marginTop="20dp"
        android:layout_gravity="center" />
    <ProgressBar
        android:id="@+id/pb"
        style="?android:attr/progressBarStyleHorizontal"
        android:layout_width="match_parent"
        android:layout_height="wrap_content"
        android:layout_marginTop="20dp"/>
</LinearLayout>
```

（2）编写 MainActivity。

```java
public class MainActivity extends AppCompatActivity {
    private Button download;
    private TextView tv;
    private ProgressBar pb;
    //定义 AsyncTask 的子类 DownloadTask
    private class DownloadTask extends AsyncTask<Integer,Integer,String>{
        @Override
        protected String doInBackground(Integer... integers) {
            for (int i=0;i<=100;i++){
                publishProgress(i);
                try {
                    Thread.sleep(integers[0]);
                }catch (InterruptedException e){
                    e.printStackTrace();
                }
            }
            return "下载完成";
        }
        @Override
        protected void onPreExecute() {
            tv.setText("");
            super.onPreExecute();
        }
        @Override
        protected void onPostExecute(String s) {
            tv.setText(s);
```

```
            super.onPostExecute(s);
        }
        @Override
        protected void onProgressUpdate(Integer... values) {
            pb.setProgress(values[0]);
            tv.setText(values[0]+"%");
            super.onProgressUpdate(values);
        }
    }
    @Override
    protected void onCreate(Bundle savedInstanceState) {
        super.onCreate(savedInstanceState);
        setContentView(R.layout.activity_main);
        download=findViewById(R.id.download);
        tv=findViewById(R.id.tv);
        pb=findViewById(R.id.pb);
        download.setOnClickListener(new View.OnClickListener() {
            @Override
            public void onClick(View v) {
                DownloadTask task=new DownloadTask();
                task.execute(100);
            }
        });
    }
}
```

运行程序，界面如图7.19（a）所示。单击"DOWNLOAD"按钮，下载过程如图7.19（b）所示，下载完成后如图7.19（c）所示。

（a）运行程序　　　　　　（b）下载过程　　　　　　（c）下载完成

图7.19　模拟下载

习题 7

1．编写一个服务类，分别采用 startService 方式和 bindService 方式启动服务，并通过 LogCat 工具观察服务生命周期方法的回调。

2．结合【例 7.1】和【例 7.2】，编写一个功能较为完善的后台音乐播放器，能够搜索 SD 卡上 Music 文件夹中的 MP3 歌曲，形成一个播放列表，并在播放界面上显示播放进度。

3．在屏幕上实时显示系统时间，如图 7.20 所示。

4．完成习题 2 中考试系统的考试倒计时功能。登录成功后进入考生界面，在该界面实时显示距离考试结束的时间，如图 7.21 所示。计时结束在屏幕上显示提示信息：考试结束。退出考试界面，如图 7.22 所示。

图 7.20　实时显示系统时间　　图 7.21　实时显示距离考试结束的时间　　图 7.22　考试结束

第 8 章　广播机制

在 Android 系统中，广播是一种应用程序之间传递消息的机制，系统级的事件都是通过广播来通知的，如网络变化、电量变化、短信接收和发送状态等。Android 系统会在发生各种系统事件时发送广播，应用程序也可以发送广播来通知其他应用程序可能感兴趣的事件（如一些新数据已下载）。为了监听这些广播信息，Android 系统提供了 BroadcastReceiver（广播接收器）组件。通过 BroadcastReceiver 可以监听系统中的广播消息，实现在不同组件之间的通信。本章介绍 Android 系统的广播机制、广播接收器的创建、系统广播的接收，以及自定义广播的发送和接收等内容。

8.1　Android 系统的广播机制

实际生活中，人们有时使用收音机来收听广播，广播电台通过特定的频率发送语音内容，只要将收音机调到广播电台指定的频率就可以收听到该电台的广播内容。在 Android 系统中，广播的发送是通过调用 sendBroadcast()方法、sendOrderedBroadcast()方法或 sendStickyBroadcast()方法来实现的，发送的广播内容是一个 Intent 对象，这个 Intent 对象可以携带要发送的数据。应用程序通过注册 BroadcastReceiver 来接收广播，只有发送广播的 Action 和接收广播的 Action 相同，广播接收者才能接收到这个广播，这和现实生活中的广播机制非常类似。

Android 系统的广播分为广播发送者和广播接收者。通常情况下，BroadcastReceiver 就是广播接收者（广播接收器）。广播发送者只负责发送广播，并不考虑接收；广播接收者需要进行注册才能接收到广播，相当于将收音机调到一定频率。Android 应用程序可以根据自己的兴趣注册广播，接收自己关心的广播内容，这些广播可能是 Android 系统发送的，也可能是其他应用程序发送的。

Android 系统中的广播使用观察者模式，基于消息的发布/订阅事件模型，实现广播发送者和接收者极大程度上的解耦，使得系统能够方便集成，更易扩展。

具体的实现流程如下。

（1）广播接收者 BroadcastReceiver 通过 Binder 机制向 AMS（Activity Manager Service）进行注册（订阅消息）；

（2）广播发送者通过 Binder 机制向 AMS 发送广播；

（3）AMS 查找符合相应条件（IntentFilter/Permission 等）的 BroadcastReceiver；

（4）AMS 将广播发送到上述符合条件的 BroadcastReceiver（一般情况下是 Activity）相应的消息循环队列中；

（5）BroadcastReceiver 通过消息循环执行收到此广播，回调 BroadcastReceiver 中的 onReceive()方法。

广播发送者和广播接收者分别属于观察者模式中位于两端的消息发布和订阅，AMS 属于中间的处理中心。发送者和接收者的执行是异步的，发出去的广播不会关心有无接收者接收，也不确定接收者到底是何时才能接收到。

Android 系统中内置了多个系统广播，在系统内部特定事件发生时，由系统自动发出，将被相应的 BroadcastReceiver 接收。

8.2 BroadcastReceiver

BroadcastReceiver 是 Android 四大基本组件之一，用于监听系统全局的广播消息，可以接收来自系统和应用程序的广播。BroadcastReceiver 本质上是一种全局的监听器，因此可以非常方便地实现系统不同组件之间的通信。

BroadcastReceiver 通过调用 onReceive()方法接收广播，并对广播通知做出反应。BroadcastReceiver 自身并不提供用户图形界面，但是当收到某个广播时，BroadcastReceiver 可以通过启动 Activity 进行响应，或者通过 Notification 来提醒用户，也可以启动 Service 等。

当 Android 系统产生一个广播事件时，可以有多个 BroadcastReceiver 接收并进行处理。广播接收器可以自由地对自己感兴趣的广播进行注册，这样当有相应的广播发出时，广播接收器就能够收到该广播，并在内部处理相应的逻辑。

8.2.1 广播接收器的创建

若想接收系统或应用程序发出的广播，先要创建广播接收器。广播接收器需要继承 BroadcastReceiver，并重写父类的 onReceive()方法。广播接收器接收到相应广播后，可自动回调 onReceive()方法。

广播接收器的创建过程与 Activity 类似。在应用程序包名处单击右键，执行"New"→"Other"→"BroadcastReceiver"菜单命令，弹出"New Android Component"对话框，如图 8.1 所示。

在 Class Name 中输入广播接收器的名称，"Exported"复选框用于选择是否允许接收当前程序之外的广播，"Enabled"复选框用于选择广播接收器是否可以由系统实例化，这两个复选框默认勾选即可，选项内容会在清单文件中显示。然后单击"Finish"按钮，广播接收器便创建完成。此时打开 MyReceiver.java，具体代码如下：

```java
public class MyReceiver extends BroadcastReceiver {
    @Override
    public void onReceive(Context context, Intent intent) {
        // TODO: This method is called when the BroadcastReceiver is receiving
        // an Intent broadcast.
        throw new UnsupportedOperationException("Not yet implemented");
    }
}
```

第 8 章 广播机制

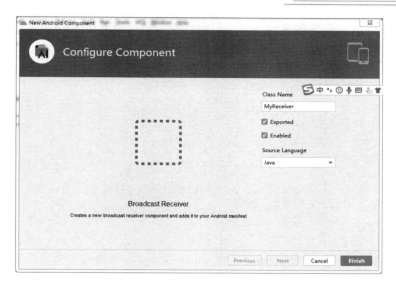

图 8.1 创建 BroadcastReceiver

在上述代码中，创建的 MyReceiver 继承自 BroadcastReceiver，默认包含 onReceive()方法，用于实现广播接收器的相关操作。一个 BroadcastReceiver 对象只有在 onReceive()方法被调用时才有效，当 onReceive()方法返回后，该对象就会被销毁，其生命周期结束。默认情况下，广播接收器运行在 UI 线程中，它的生命周期只有 10 秒钟左右，如果 onReceive()方法不能在这几秒内完成操作，Android 系统就会认为该 BroadcastReceiver 对象无响应，将会报 ANR（Application Not Response）错误信息，因此，onReceive()方法中不能执行耗时的操作。如果在收到广播后需要处理耗时的业务，应该开启 Service 进行处理，而不应该在 BroadcastReceiver 开启新线程来完成耗时操作，因为 BroadcastReceiver 的生命周期很短，子线程可能还没有结束 BroadcastReceiver 就先结束了。当 BroadcastReceiver 结束时，其所在的进程就属于空进程，没有任何活动组件的进程，在系统需要内存时容易被优先杀死。如果 BroadcastReceiver 的宿主进程被杀死，那么正在工作的子线程也会被一起杀死，因此采用子线程是不可靠的。一般情况下，根据实际业务需求，onReceive()方法中都会涉及与其他组件之间的交互，如发送 Notification、启动 Service 等。

8.2.2 广播接收器的注册

创建的广播接收器必须经过注册才能接收相应的广播事件，注册的方式有两种：静态注册和动态注册。

1．静态注册

创建广播接收器之后，在 AndroidManifest.xml 文件中对其进行注册。使用 Android Studio 向导创建广播接收器后，在 application 标签中出现了一个新的标签<receiver>。

```
        <receiver
            android:name=".MyReceiver"
            android:enabled="true"
            android:exported="true">
        </receiver>
```

但是，此时注册的广播接收器还不能接收广播，必须为其设置<intent-filter>过滤器，指定其可以接收的广播类型。例如，当系统收到短信时，会发出一个 Action 名称为"android.provier.Telephony.SMS_RECEIVED"的广播，如果广播接收器 MyReceiver 要接收该广播，则必须在其<intent-filter>中添加广播的 Action，代码如下：

```xml
<receiver
    android:name=".MyReceiver"
    android:enabled="true"
    android:exported="true">
    <intent-filter>
        <action android:name="android.provier.Telephony.SMS_RECEIVED" />
    </intent-filter>
    ...
</receiver>
```

静态注册方式由系统来管理 BroadcastReceiver 对象。当系统或其他应用程序发出广播时，Android 系统的包管理器就会检查所有已安装的包中的配置文件有没有匹配的 Action，如果有且可以接收，那么就调用这个 BroadcastReceiver 对象，然后执行 OnReceive()方法，当 OnReceive()方法执行完之后，BroadcastReceiver 对象就会被销毁。BroadcastReceiver 对象是在 Intent 匹配后再实例化的，而且每次都是重新实例化。

2．动态注册

在 Activity 中实例化定义的广播接收器、实例化过滤器，设置要过滤的广播类型，并调用 Context 的 registerReceiver(BroadcastReceiver receiver, IntentFilter filter)方法注册广播，其中，参数 receiver 是广播接收器对象，filter 是过滤器对象。在此 Activity 结束时调用 unregisterReceiver(BroadcastReceiver receiver) 方法注销该广播。

例如，广播接收器 MyReceiver 要接收 android.provier.Telephony.SMS_RECEIVED 广播，动态注册代码如下：

```java
MyReceiver receiver;
@Override
protected void onCreate(Bundle savedInstanceState) {
    super.onCreate(savedInstanceState);
    setContentView(R.layout.activity_main);
    receiver=new MyReceiver();
    IntentFilter filter = new IntentFilter();
    filter.addAction("android.provider.Telephony.SMS_RECEIVED");
    context.registerReceiver(receiver,filter);
}
@Override
protected void onDestroy() {
    super.onDestroy();
    unregisterReceiver(receiver);
}
```

在 Activity 中注册广播，当 Activity 实例化时，就会动态地将 MyReceiver 注册到系统中。在此 Activity 销毁时,广播接收器被注销,将不再接收相应的广播。一般在 Activity 的 onStart ()方法中注册广播，在 onDestroy ()方法中注销广播，广播接收器则跟随 Activity 的生命周期。

8.3 接收系统广播

Android 系统中内置了多个系统广播,当系统内部发生特定事件时,由系统自动发出。每个系统广播都具有特定的 intent-filter,其中包括具体的 Action,系统广播发出后,将被相应的 BroadcastReceiver 接收。广播消息被封装在一个 Intent 对象中,该对象的操作字符串标识所发生的事件。用户若要接收系统广播,只需创建并注册相应的广播接收器即可。

【例 8.1】监听手机锁屏与解锁。

手机锁屏时系统会发出 android.intent.action.SCREEN_OFF 广播,当屏幕唤醒时系统会发出 android.intent.action.SCREEN_ON 广播,这两个广播只能通过动态注册。

(1) 创建应用程序,自定义广播接收器 MyReceiver。

```java
public class MyReceiver extends BroadcastReceiver {
    @Override
    public void onReceive(Context context, Intent intent) {
        //获取到当前广播的事件类型
        String action = intent.getAction();
            //判断当前广播的事件类型
        if(action.equals(Intent.ACTION_SCREEN_ON)) {
            Log.i("TAG", "屏幕锁屏了");
        } else if (action.equals(Intent.ACTION_SCREEN_OFF)) {
            Log.i("TAG" , "屏幕解锁了");
        }
    }
}
```

(2) 动态注册广播。

```java
public class MainActivity extends AppCompatActivity {
    MyReceiver receiver;
    @Override
    protected void onCreate(Bundle savedInstanceState) {
        super.onCreate(savedInstanceState);
        setContentView(R.layout.activity_main);
        receiver=new MyReceiver();     //创建广播接收器对象
        IntentFilter filter=new IntentFilter();    //创建过滤器对象
        filter.addAction(Intent.ACTION_SCREEN_ON);   //要监听的广播 Action
        filter.addAction(Intent.ACTION_SCREEN_OFF);  //要监听的广播 Action
        registerReceiver(receiver, filter);   //注册广播
    }
    @Override
    protected void onDestroy() {
        super.onDestroy();
        unregisterReceiver(receiver);    //注销广播
    }
}
```

运行程序,多次按手机右侧的锁屏键,日志输出窗口的信息如图 8.2 所示。

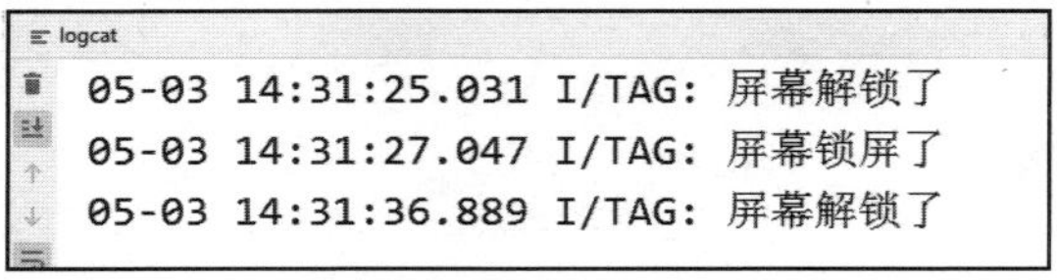

```
05-03 14:31:25.031 I/TAG: 屏幕解锁了
05-03 14:31:27.047 I/TAG: 屏幕锁屏了
05-03 14:31:36.889 I/TAG: 屏幕解锁了
```

图 8.2 锁屏与解锁的日志

【例 8.2】监听网络状态变化。

当手机的网络状态发生改变时，系统会发送一个 Action 为 android.net.conn.CONNECTIVITY_CHANGE 的广播。在 Android 7.0（API 24）及以上版本，监听 CONNECTIVITY_ACTION 广播需要动态注册。

另外，访问系统的网络状态需要相应权限，可在 AndroidManifest.xml 文件中加入访问系统网络状态的权限：

```xml
<uses-permission android:name="android.permission.ACCESS_NETWORK_STATE" />
```

利用 getSystemService 可以得到专门用于管理网络连接的系统服务类 ConnectivityManager，然后利用它的 getActiveNetworkInfo() 得到一个 NetworkInfo 的实例，并利用该实例判断当前网络的状态。

（1）定义 NetworkChangedReceiver 类。

```java
public class NetworkChangedReceiver extends BroadcastReceiver {
    @Override
    public void onReceive(Context context, Intent intent) {
        //获取管理网络连接的系统服务类的实例 connectivityManager
        ConnectivityManager connectivityManager = (ConnectivityManager)
                    getSystemService(Context.CONNECTIVITY_SERVICE);
        //NetworkInfo 对象包含网络连接的所有信息
        NetworkInfo networkInfo=connectivityManager.getActiveNetworkInfo();
        //调用 NetworkInfo 的 isAvailable() 方法判断是否联网
        if(networkInfo!=null&&networkInfo.isAvailable()) {
            Log.i("TAG", "网络已连接");
        }else {
            Log.i("TAG", "网络不可用");
        }
    }
}
```

（2）注册广播接收器。

在 MainActivity 的 onCreate() 方法中注册广播接收器。

```java
public class MainActivity extends AppCompatActivity {
    private IntentFilter filter;
    private NetworkChangedReceiver receiver;
    @Override
    protected void onCreate(Bundle savedInstanceState) {
        super.onCreate(savedInstanceState);
        setContentView(R.layout.activity_main);
        filter = new IntentFilter();
        //为过滤器添加处理规则
        filter.addAction("android.net.conn.CONNECTIVITY_CHANGE");
        receiver = new NetworkChangedReceiver();
```

```
            //注册广播接收器
            registerReceiver(receiver, filter);
    }
        @Override
        protected void onDestroy() {
        super.onDestroy();
        //动态的广播接收器最后一定要取消注册
        unregisterReceiver(receiver);
    }
```

运行程序，改变模拟器的网络状态，LogCat 窗口的日志信息如图 8.3 所示。

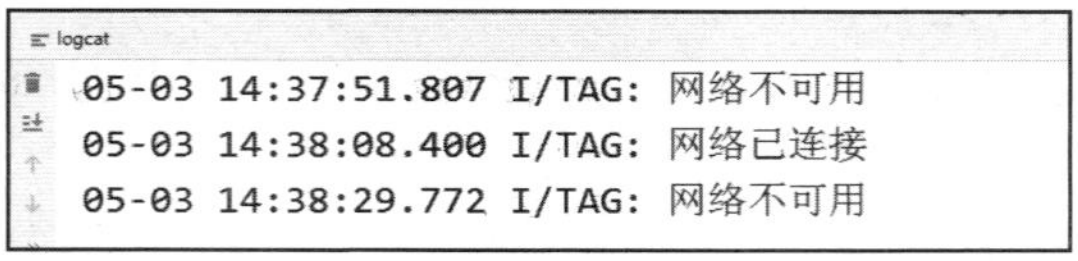

```
05-03 14:37:51.807 I/TAG: 网络不可用
05-03 14:38:08.400 I/TAG: 网络已连接
05-03 14:38:29.772 I/TAG: 网络不可用
```

图 8.3　网络状态信息

8.4　自定义广播

8.4.1　广播类型

在实际开发中，有时为了满足一些特殊需求需要自定义广播。Android 应用程序不仅可以接收系统广播，还可以接收其他应用程序发出的广播，或者向其他应用程序发送广播。

Android 中发送广播有两种形式。

1．普通广播（Normal broadcasts）

普通广播是一种完全异步执行的广播，在广播发出之后，所有的广播接收器几乎会在同一时刻接收到这条广播消息。这种广播消息传递的效率比较高，但缺点是接收者不能将处理结果传递给下一个接收者，并且广播无法被拦截。

Context 类提供了以下方法用于发送普通广播：

```
sendBroadcast(Intent intent);
sendBroadcast(Intent intent, String receiverPermission);
sendBroadcastAsUser(Intent intent, UserHandle user);
sendBroadcastAsUser(Intent intent, UserHandle user, String receiverPermission);
```

2．有序广播（Ordered broadcasts）

有序广播是一种同步执行的广播，在广播发出之后，同一时刻只会有一个广播接收器能够收到这条广播消息，当这个广播接收器中的逻辑执行完毕之后，广播才会继续传递。

有序广播的接收是有先后顺序的，系统根据接收者声明的优先级按顺序逐个执行接收者，优先级高的广播接收者先接收广播。优先级高的接收者可以对广播进行修改，通过 setResultData()方法和 setResultExtras()方法将处理的结果存入 Intent，传递给下一个接收者。然后，下一个接收者通过 getResultData()和 getResultExtras(true)接收高优先级的接收者存入的数据。优先级高的接收者也有权使用 abortBroadcast()拦截广播，使后面的接收者无法接收到该条广播。

相比无序广播，有序广播的效率较低，但此类广播的接收是有先后顺序的，并可被拦截。有序广播的定义过程与普通广播无异，只是其发送方式变为 sendOrderedBroadcast()方法。

静态注册广播接收器时，可在<intent-filter>标签中使用 priority 属性设置优先级，数值越大优先级别越高。动态注册广播接收器时优先级可通过 filter.setPriority()方法设置。

如果两个广播接收器的优先级相同，则先注册的优先级别高。

8.4.2 普通广播

【例 8.3】普通广播的发送与接收。

本例演示应用程序之间普通广播的发送和接收，一个应用程序负责发送广播，另一个应用程序负责接收广播。

（1）创建应用程序 ForHelp。

该应用程序用于发送广播，单击界面上的"呼救"按钮，发送一条广播，模拟危险情况下发出的求救信号。该应用程序的 MainActivity.java 代码如下：

```java
public class MainActivity extends AppCompatActivity {
    private Button button;
    @Override
    protected void onCreate(Bundle savedInstanceState) {
        super.onCreate(savedInstanceState);
        setContentView(R.layout.activity_main);
        button=findViewById(R.id.button);
        button.setOnClickListener(new View.OnClickListener() {
            @Override
            public void onClick(View v) {
                Intent intent=new Intent();
                intent.setAction("Help");    //广播类型
                intent.putExtra("info","help！help...");
                sendBroadcast(intent);    //发送广播
            }
        });
    }
}
```

（2）创建应用程序 HelpCenter。

在 HelpCenter 中创建广播接收器 HelpReceiver，用于接收 ForHelp 应用程序发出的广播。

```java
public class HelpReceiver extends BroadcastReceiver {
    @Override
    public void onReceive(Context context, Intent intent) {
        Log.i("msg"," HelpReceiver 收到求救信息："+intent.getStringExtra("info"));
    }
}
```

（3）注册广播接收器。

在应用程序 HelpCenter 的清单文件中注册广播接收器 HelpReceiver。

```xml
<receiver
    android:name=".HelpReceiver"
    android:enabled="true"
```

```
                android:exported="true" >
                <!--接收的广播类型为字符串 Help-->
                <intent-filter>
                    <action android:name="Help"/>
                </intent-filter>
            </receiver>
```

运行应用程序 HelpCenter 和 ForHelp，单击 ForHelp 界面的"呼救"按钮，LogCat 窗口输出的日志信息如图 8.4 所示。

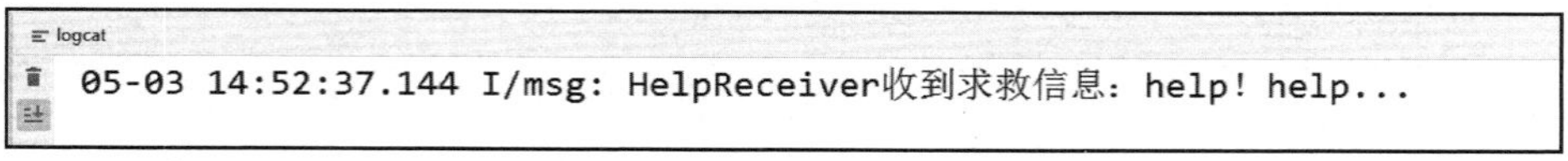

```
logcat
05-03 14:52:37.144 I/msg: HelpReceiver收到求救信息: help! help...
```

图 8.4　广播信息

8.4.3　有序广播

【例 8.4】有序广播的发送与接收。

本例演示应用程序之间有序广播的发送和接收，因此需要创建多个广播接收器来接收广播。

（1）发送有序广播。

将应用程序 ForHelp 的 MainActivity.java 代码中的语句 sendBroadcast(intent);改为 sendOrderedBroadcast(intent,null);使之发送有序广播。

（2）创建两个广播接收器。

在应用程序 HelpCenter 中，再创建两个广播接收器 SecondHelpReceiver 和 ThirdHelpReceiver。

```java
public class SecondHelpReceiver extends BroadcastReceiver {
    @Override
    public void onReceive(Context context, Intent intent) {
        Log.i("msg","SecondHelpReceiver 接收到求救信息："+intent.getStringExtra("info"));
    }
}
public class ThirdHelpReceiver extends BroadcastReceiver {
    @Override
    public void onReceive(Context context, Intent intent) {
        Log.i("msg","ThirdHelpReceiver 接收到求救信息："+intent.getStringExtra("info"));
    }
}
```

（3）注册广播接收器并设置优先级。

在 HelpReceiver 的清单文件中设置各个广播接收器的优先级，代码如下：

```xml
<receiver
    android:name=".HelpReceiver"
    android:enabled="true"
    android:exported="true">
    <!--接收的广播类型为字符串 Help-->
    <intent-filter android:priority="1000">
        <action android:name="Help"/>
    </intent-filter>
```

```xml
        </receiver>
        <receiver
            android:name=".SecondHelpReceiver"
            android:enabled="true"
            android:exported="true" >
                <intent-filter android:priority="1100">
                    <action android:name="Help"/>
                </intent-filter>
        </receiver>
        <receiver
            android:name=".ThirdHelpReceiver"
            android:enabled="true"
            android:exported="true">
                <intent-filter android:priority="1200">
                    <action android:name="Help"/>
                </intent-filter>
        </receiver>
```

运行应用程序 HelpCenter 和 ForHelp，单击 ForHelp 界面的"呼救"按钮，LogCat 窗口输出的日志信息如图 8.5 所示。

```
logcat
05-03 15:04:58.923 I/msg: ThirdHelpReceiver接收到求救信息: help! help...
05-03 15:04:58.928 I/msg: SecondHelpReceiver接收到求救信息: help! help...
05-03 15:04:58.930 I/msg: HelpReceiver收到求救信息: help! help...
```

图 8.5 广播信息

首先接收到的广播是优先级最高的 ThirdHelpReceiver，其次是 SecondHelpReceiver，最后是 HelpReceiver，由此可以看出，广播接收器的优先级决定了接收广播的先后顺序。

接下来，在 SecondHelpReceiver 中通过 abortBroadcast()方法来拦截广播，其他优先级低的接收器则无法接收该广播。

```java
public class SecondHelpReceiver extends BroadcastReceiver {
    @Override
    public void onReceive(Context context, Intent intent) {
        Log.i("msg","SecondHelpReceiver 接收到求救信息: "+intent.getStringExtra("info"));
        abortBroadcast();    //拦截广播
    }
}
```

先重新运行应用程序 HelpReceiver，然后单击 ForHelp 界面的"呼救"按钮，LogCat 窗口输出的日志信息如图 8.6 所示。

```
logcat
05-03 15:13:57.652 I/msg: ThirdHelpReceiver接收到求救信息: help! help...
05-03 15:13:57.653 I/msg: SecondHelpReceiver接收到求救信息: help! help...
```

图 8.6 拦截广播

可以看出，广播接收器 SecondHelpReceiver 成功拦截了广播，优先级最低的广播接收器 HelpReceiver 则无法接收该广播。注意，只有有序广播才能被拦截。

上述广播接收器是在清单文件中静态注册的，如果采用动态注册，代码如下：

```java
public class MainActivity extends AppCompatActivity {
    private HelpReceiver helpReceiver;
    private SecondHelpReceiver secondHelpReceiver;
    private ThirdHelpReceiver thirdHelpReceiver;
    @Override
    protected void onCreate(Bundle savedInstanceState) {
        super.onCreate(savedInstanceState);
        setContentView(R.layout.activity_main);
        helpReceiver=new HelpReceiver();
        IntentFilter filter=new IntentFilter();
        filter.addAction("Help");
        filter.setPriority(1000);
        registerReceiver(helpReceiver,filter);
        secondHelpReceiver=new SecondHelpReceiver();
        IntentFilter secondFilter=new IntentFilter();
        secondFilter.addAction("Help");
        secondFilter.setPriority(1100);
        registerReceiver(secondHelpReceiver,secondFilter);
        thirdHelpReceiver=new ThirdHelpReceiver();
        IntentFilter thirdFilter=new IntentFilter();
        thirdFilter.addAction("Help");
        thirdFilter.setPriority(1200);
        registerReceiver(thirdHelpReceiver,thirdFilter);
    }
    @Override
    protected void onDestroy() {
        super.onDestroy();
        unregisterReceiver(helpReceiver);
        unregisterReceiver(secondHelpReceiver);
        unregisterReceiver(thirdHelpReceiver);
    }
}
```

8.5 本地广播

前面所讲的广播属于全局广播，可以被其他任何程序接收，这样就很容易引起安全性问题。例如，发送一些携带关键性数据的广播时就有可能被其他的应用程序截获，或者其他的程序不停地向广播接收器发送各种垃圾广播。

为了解决上述安全性问题，Android 引入了本地广播机制。本地广播只能够在应用程序的内部进行传递，并且广播接收器也只能接收来自本应用程序发出的广播，从而保证了安全性。

本地广播的用法并不复杂，主要使用 LocalBroadcastManager 来对广播进行管理，并提供发送广播和注册广播接收器的方法。

首先通过 LocalBroadcastManager 的 getInstance()方法得到一个实例，然后在注册广播接收器时调用 LocalBroadcastManager 实例的 registerReceiver()方法，在发送广播时调用 LocalBroadcastManager 实例的 sendBroadcast() 方法。

使用 LocalBroadcastManager 和使用全局广播差不多,都需要先注册再发送广播,最后注销注册。

```
localBroadcastManager = LocalBroadcastManager.getInstance(this);
localBroadcastManager.registerReceiver(mBroadcastReceiver, intentFilter);
```

注意,本地广播是无法通过静态注册的方式来接收的。

习题 8

1. 编写程序,实现两个 App 之间通过自定义广播进行数据通信。
2. 编写程序,通过 BroadcastReceiver 实现在插拔耳机时让用户调整音量的提示。
3. 编写 Android 广播接收器,实现拦截打出电话时系统发出的广播。

第 9 章 网络编程

在移动互联网时代，越来越多的 Android App 需要使用互联网才能够提供更多的功能，因此，网络编程在 Android 开发中的地位也变得越来越重要。本章介绍 Android 网络编程的相关知识。

9.1 使用 HTTP 访问网络

网络编程主要是指通过程序对网络设备之间的通信进行控制，网络通信可以在手机等移动设备之间进行，也可以在移动设备与网络服务器之间进行，如图 9.1 所示。

为确保不同架构的设备之间通信能正常进行，既可以使用无线局域网 TCP/UPD 套接字通信，也可以使用移动互联网 HTTP 通信。

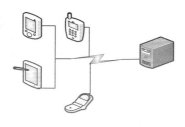

图 9.1 网络连接

9.1.1 网络编程基本概念

1．网络架构

网络架构主要有两种模式 B/S 和 C/S。其中，B/S 模型是浏览器/服务器端模式，通过应用层的 HTTP 协议通信，不需要特定客户端软件，而是需要统一规范的客户端，简而言之就是 Android 网络浏览器访问 Web 服务器端的方式；C/S 模型是客户端/服务器端模式，通过任意的网络协议通信，需要特定的客户端软件。

2．HTTP 通信协议

HTTP（Hypertext Transfer Protocol，超文本传输协议）是 TCP/IP 的一个应用层协议，用于定义 Web 浏览器与 Web 服务器之间交换数据的过程。客户端连上 Web 服务器后，若想获得 Web 服务器中的某个 Web 资源，需遵守一定的通信格式，HTTP 用于定义客户端与 Web 服务器通信的格式。底层工作原理是通过三次握手获得网络链路，完成数据传输和断开连接，具体过程如下。

第一次握手：客户端发送 SYN 包（SYN=j）到服务器，并进入 SYN_SEND 状态，等待服务器确认；

第二次握手：服务器收到 SYN 包，必须确认客户的 SYN（ACK=j+1），同时自己也发送一个 SYN 包（SYN=k），即 SYN+ACK 包，此时服务器进入 SYN_RECV 状态；

第三次握手：客户端收到服务器的 SYN＋ACK 包，向服务器发送确认包 ACK（ACK=

k+1），此包发送完毕，客户端和服务器进入 ESTABLISHED 状态，完成三次握手。握手过程中传送的包里不包含数据，三次握手完毕后，客户端与服务器才能正式开始传送数据。理想状态下，TCP 连接一旦建立，在通信双方中的任何一方主动关闭连接之前，TCP 连接都将被一直保持下去。断开连接时服务器和客户端均可以主动发起断开 TCP 连接的请求，断开过程需要经过"四次握手"，第一次握手：客户端发送报文告诉服务器没有数据要发送了；第二次握手：服务器收到，再发送给客户端告诉它已收到；第三次握手：服务器向客户端发送报文，请求关闭连接；第四次握手：客户端收到关闭连接的请求，向服务器发送报文，服务器关闭连接，如图 9.2 所示。

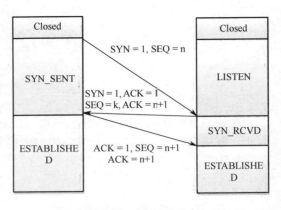

图 9.2　三次握手示意

HTTP URL 的格式为 http://host[":"port][abs_path]，其中，http 表示要通过 HTTP 来定位网络资源；host 表示合法的 Internet 主机域名或 IP 地址；port 表示指定一个端口号，若为空则使用默认端口 80；abs_path 指定请求资源的 URI（Web 上任意的可用资源）。

HTTP 的两种报文分别是请求报文和响应报文。HTTP 请求报文由请求行、请求报头、空行和请求数据四部分组成。HTTP 请求的方法有八种，分别是 GET、POST、DELETE、PUT、HEAD、TRACE、CONNECT、OPTIONS，其中 PUT、DELETE、POST、GET 分别对应增、删、改、查的操作，对于移动开发最常用的就是 POST 和 GET。HTTP 的响应报文由状态行、消息报头、空行、响应正文组成。在通信完成后应用程序根据状态行中的状态码来判断通信完成的情况，并以此为判断依据进行下一步工作。状态代码由三位数字组成，第一个数字定义了响应的类别，共有五种可能的取值。

- 100～199：指示信息，表示请求已接收，可继续处理。
- 200～299：请求成功，表示请求已被成功接收、理解。
- 300～399：重定向，要完成请求必须进行更进一步的操作。
- 400～499：客户端错误，请求有语法错误或请求无法实现。
- 500～599：服务器端错误，服务器未能实现合法的请求。

常见的状态码如下。

- 200 OK：客户端请求成功。
- 400 Bad Request：客户端请求有语法错误，不能被服务器所理解。
- 401 Unauthorized：请求未经授权，这个状态代码必须和 WWW-Authenticate 报头域一起使用。
- 403 Forbidden：服务器收到请求，但是拒绝提供服务。
- 500 Internal Server Error：服务器发生不可预期的错误。
- 503 Server Unavailable：服务器当前不能处理客户端的请求，一段时间后可能恢复正常。

9.1.2　使用 HttpURLConnection 连接网络

Android 应用程序可以使用 GET 方法和 POST 方法发送 HTTP 网络请求，一个完整的

HTTP 请求需要经历两个过程：客户端发送请求到服务器，然后服务器将结果返回给客户端，如图 9.3 所示。

Android 团队从 Android API Level 9 开始实现了一个发送 HTTP 请求的客户端类——HttpURLConnection，它是一种多用途、轻量级的 HTTP 客户端，HttpURLConnection 是基于 HTTP 的，支持 GET、POST、PUT、DELETE 等各种请求方式，使用它来进行 HTTP 操作可以适用于大多数的应用程序。使用 HttpURLConnection 对象进行网络连接的过程如下。

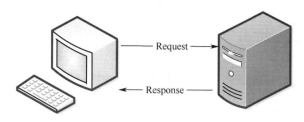

图 9.3　HTTP 网络请求

（1）创建一个 URL 对象。

```
URL url = new URL("http://www.baidu.com");
```

（2）利用 HttpURLConnection 对象从网络中获取网页数据。

```
HttpURLConnection conn = (HttpURLConnection) url.openConnection();
```

（3）设置连接超时。

```
conn.setConnectTimeout(6*1000);
```

（4）对响应码进行判断。

```
if (conn.getResponseCode() != 200)    //从 Internet 获取网页，发送请求，将网页以流的形式读回来
    throw new RuntimeException("请求 url 失败");
```

（5）获得网络返回的输入流。

```
InputStream is = conn.getInputStream();
String result = readData(is, "UTF-8");
conn.disconnect();
```

【例 9.1】HttpURLConnection 基本用法示例。

（1）创建应用程序，编辑主界面布局。

创建一个应用程序，编辑 XML 布局文件，使用 LinearLayout 布局容器，在其中添加一个 Button 控件和一个 ImageView 控件，单击按钮程序从指定 URL 下载图片内容，并显示在 ImageVIew 控件中，代码如下：

```xml
<?xml version="1.0" encoding="utf-8"?>
<LinearLayout
    xmlns:android="http://schemas.android.com/apk/res/android"
    xmlns:app="http://schemas.android.com/apk/res-auto"
    xmlns:tools="http://schemas.android.com/tools"
    android:layout_width="match_parent"
    android:layout_height="match_parent"
    android:orientation="vertical"
    tools:context=".chap10.s1.HttpUrlConnectionActivity">
    <Button
        android:id="@+id/btn1"
        android:layout_width="wrap_content"
        android:layout_height="wrap_content"
        android:text="获取图片" />
    <ImageView
        android:id="@+id/img1"
        android:layout_width="200dp"
```

```
            android:layout_height="200dp"
            android:src="@mipmap/ic_launcher" />
</LinearLayout>
```

（2）创建 Web 服务器。

本例使用 Tomcat 软件创建一个名称为"AndroidWeb"的 Web 服务器端程序，将图片文件"e.jpg"复制到服务器根目录中。

（3）因为需要访问网络，所以要在"AndroidManeifest.xml"文件中添加权限申请代码。

```
<uses-permission android:name="android.permission.INTERNET" />
```

同时在"Application"标签中添加属性：

```
android:usesCleartextTraffic="true"
```

（4）编辑 Activity 类代码。

首先，在 onCreate 函数内初始化控件对象：

```
btn1 = (Button) findViewById(R.id.btn1);
img1 = (ImageView) findViewById(R.id.img1);
```

其次，完成下载任务代码。访问网络，下载资源是一个耗时操作，因此使用 AsyncTask 类进行并发操作，代码如下：

```
class DownImgAsyncTask extends AsyncTask<String, Void, Bitmap> {
    @Override
    protected void onPreExecute() {
        super.onPreExecute();
        img1.setImageBitmap(null);
    }
    @Override
    protected void onPostExecute(Bitmap result) {
        super.onPostExecute(result);
        if (result != null) {
            img1.setImageBitmap(result);
        }
    }
    @Override
    protected Bitmap doInBackground(String... params) {
        Bitmap b = getImageBitmap(params[0]);
        return b;
    }
}
```

AsyncTask 类需要三个参数，分别代表传入参数、进度对象和返回值类型，关键函数是 onPreExecute、onPostExecute 和 doInBackground，它们分别执行异步的前、后操作和异步操作本身，在本示例中异步操作就是下载图片文件，代码如下：

```
private Bitmap getImageBitmap(String url) {
    URL imgurl = null;
    Bitmap bitmap = null;
    HttpURLConnection urlConnection;
    try {
        imgurl = new URL(url);
        urlConnection = (HttpURLConnection) imgurl.openConnection();
        urlConnection.setRequestMethod("GET");
        urlConnection.connect();
```

```
            InputStream is = urlConnection.getInputStream();
            bitmap = BitmapFactory.decodeStream(is);
            is.close();
        } catch (IOException e) {
            e.printStackTrace();
        }
        return bitmap;
    }
```

使用图片文件的 URL 创建一个 HttpURLConnection 对象，并通过该对象获取输入流，这个输入流就是图片文件的内容，然后以此流为参数获取 BitMap 对象。至此，可以为按钮对象设置事件监听器，在事件处理函数中启动下载异步操作，代码如下：

```
btn1.setOnClickListener(new View.OnClickListener() {
    @Override
    public void onClick(View v) {
        String imgUrl = "http://192.168.43.33:8080/AndroidWeb/e.jpg";
        new DownImgAsyncTask().execute(imgUrl);
    }
});
```

图 9.4　使用 HttpURLConnection 下载图片

（5）运行程序

单击"获取图片"按钮，可以看到图片视图控件的内容发生了改变，程序运行效果如图 9.4 所示。

9.1.3　网络信息传输

HttpURLConnection 使用 GET 方式访问网络时，如果需要传递参数，则直接把参数拼接到 url 后面，代码如下：

```
String url = "https://(域名:端口)?userName=zhangsan&password=123456";
```

要求是："url"与参数之间用"？"隔开；键值对中键与值用"="连接；两个键值对之间用"&"连接。

当使用 POST 方式向 Internet 发送请求参数时，其步骤如下。

（1）创建 URL 对象。

```
URL realUrl = new URL(requestUrl);
```

（2）通过 HttpURLConnection 对象向网络地址发送请求。

```
HttpURLConnection conn = (HttpURLConnection) realUrl.openConnection();
```

（3）设置容许输出。

```
conn.setDoOutput(true);
```

（4）设置不使用缓存。

```
conn.setUseCaches(false);
```

（5）设置使用 POST 的方式发送。

```
conn.setRequestMethod("POST");
```

（6）设置维持长连接。

```
conn.setRequestProperty("Connection", "Keep-Alive");
```

（7）设置文件字符集。
```
conn.setRequestProperty("Charset", "UTF-8");
```
（8）设置文件长度。
```
conn.setRequestProperty("Content-Length",String.valueOf(data.length));
```
（9）设置传输内容类型属性。
```
conn.setRequestProperty("Content-Type","application/x-www-form-urlencoded");
```
（10）以流的方式向服务器输出数据。

在 POST 方式下传输数据时需要注意的是，发送 POST 请求必须设置允许输出；不要使用缓存，否则容易出现问题；用 HttpURLConnection 对象的 setRequestProperty()进行设置生成 HTML 文件头。

【例 9.2】POST 传递参数示例。

（1）创建应用程序，编辑主界面布局。

创建一个应用程序，编辑 XML 布局文件，实现一个登录界面，在输入框内填写账户和密码，单击"登录"按钮完成登录操作，代码如下：

```xml
<?xml version="1.0" encoding="utf-8"?>
<LinearLayout xmlns:android="http://schemas.android.com/apk/res/android"
    xmlns:app="http://schemas.android.com/apk/res-auto"
    xmlns:tools="http://schemas.android.com/tools"
    android:layout_width="match_parent"
    android:layout_height="match_parent"
    android:orientation="vertical"
    tools:context=".chap10.s1.PostActivity">
    <LinearLayout
        android:layout_width="match_parent"
        android:layout_height="wrap_content">
        <TextView
            android:layout_width="wrap_content"
            android:layout_height="wrap_content"
            android:text="账户" />
        <EditText
            android:id="@+id/name"
            android:layout_width="match_parent"
            android:layout_height="wrap_content" />
    </LinearLayout>
    <LinearLayout
        android:layout_width="match_parent"
        android:layout_height="wrap_content">
        <TextView
            android:layout_width="wrap_content"
            android:layout_height="wrap_content"
            android:text="密码" />
        <EditText
            android:id="@+id/password"
            android:layout_width="match_parent"
            android:layout_height="wrap_content" />
    </LinearLayout>
    <Button
```

```xml
            android:id="@+id/btnLogin"
            android:layout_width="wrap_content"
            android:layout_height="wrap_content"
            android:text="登录" />
</LinearLayout>
```

（2）编辑 Activity 类代码。

在 onCreate 函数内初始化控件对象，为按钮对象添加事件监听器，并在事件处理函数中以 POST 方式向服务器发送用户的输入值。由于网络操作是一个耗时操作，所以在这里开启了一个工作线程，代码如下：

```java
loginBtn.setOnClickListener(new View.OnClickListener() {
    @Override
    public void onClick(View view) {
        new Thread() {
            public void run() {
                try {
                    HttpURLConnection conn = null;
                    String strUrl = "http://192.168.43.33:8080/
                        AndroidWeb/GetPostServlet";
                    conn = (HttpURLConnection) new URL(strUrl)
                        .openConnection();
                    conn.setRequestMethod("POST");
                    conn.setReadTimeout(5000);
                    conn.setConnectTimeout(5000);
                    conn.setDoOutput(true);
                    conn.setDoInput(true);
                    conn.setUseCaches(false);
                    String data="password="+URLEncoder.encode(
                        password.getText().toString(),
                        "UTF-8") + "&userName="+
                        URLEncoder.encode(
                        userName.getText().toString(), "UTF-8");
                    OutputStream out = conn.getOutputStream();
                    out.write(data.getBytes());
                    out.flush();
                    if (conn.getResponseCode() == 200) {
                        InputStream is = conn.getInputStream();
                        ByteArrayOutputStream message = new
                                ByteArrayOutputStream();
                        int len = 0;
                        byte buffer[] = new byte[1024];
                        while ((len = is.read(buffer)) != -1) {
                            message.write(buffer, 0, len);
                        }
                        is.close();
                        message.close();
                        result = new String(message.toByteArray());
                        handler.sendEmptyMessage(0x123);
                    }
                } catch (IOException e) {
                    e.printStackTrace();
```

```
                    }
                }
            }.start();
        }
    });
```

通过服务器端获取用户名和密码,判断账户是否合法,并返回结果。这里使用 Servlet 技术和 doPost 函数处理上述逻辑,代码如下:

```
@Override
protected void doPost(HttpServletRequest req,
    HttpServletResponse resp) throws ServletException, IOException {
        String userName = req.getParameter("userName");
        String password = req.getParameter("password");
        if ("test".equals(userName) && "123456".equals(password)) {
            resp.getWriter().print("Login OK...");
        } else {
            resp.getWriter().print("Login error...");
        }
    }
```

客户端获取服务器的处理结果并在界面上显示,因为,Android 的界面元素不是线程安全的,即在工作线程中无法改变控件状态,所以这里使用 Handler 技术异步更新,代码如下:

```
private Handler handler = new Handler() {
    public void handleMessage(android.os.Message msg) {
        Toast.makeText(PostActivity.this, result, Toast.LENGTH_SHORT)
            .show();
    }
};
```

(3)运行程序。

输入账号"test"和密码"123456",单击"登录"按钮显示登录成功结果,效果如图 9.5 所示。

图 9.5 运行程序界面

9.1.4 XML 和 JSON

在网络中，数据交互通常是以 XML 和 JSON 的格式进行，这里分别进行介绍。

1．向 Internet 发送 XML 数据

XML 格式是通信的标准语言，Android 系统也可以通过发送 XML 文件传输数据。使用 HttpURLConnector 发送 XML 数据的步骤如下：

（1）将生成的 XML 文件写入 byte 数组中，并设置字符编码集为"UTF-8"。
```
byte[] xmlbyte = xml.toString().getBytes("UTF-8");
```
（2）创建 URL 对象，并指定地址和参数。
```
URL url = new URL(http://"域名":"端口"/"应用"/"路径");
```
（3）获得链接。
```
HttpURLConnection conn = (HttpURLConnection) url.openConnection();
```
（4）设置连接超时。
```
conn.setConnectTimeout(6* 1000);
```
（5）设置允许输出。
```
conn.setDoOutput(true);
```
（6）设置不使用缓存。
```
conn.setUseCaches(false);
```
（7）设置以 POST 方式传输。
```
conn.setRequestMethod("POST");
```
（8）维持长连接。
```
conn.setRequestProperty("Connection", "Keep-Alive");
```
（9）设置字符集。
```
conn.setRequestProperty("Charset", "UTF-8");
```
（10）设置文件的总长度。
```
conn.setRequestProperty("Content-Length", String.valueOf(xmlbyte.length));
```
（11）设置文件类型。
```
conn.setRequestProperty("Content-Type", "text/xml; charset=UTF-8");
```
（12）以文件流的方式发送 XML 数据。
```
outStream.write(xmlbyte);
```

可以看出使用 HttpURLConnector 创建网络连接传输 XML 数据也是流操作，同样需要遵循 HTTP 的要求，即对传输文件的大小有所限制，一般在 5MB 以下。

Android 提供了三种解析 XML 的方式，分别是 SAX（Simple API XML）、DOM（Document Object Model）和 PULL。其中，PULL 解析器小巧轻便，解析速度快，简单易用，非常适合在 Android 移动设备中使用，Android 系统内部在解析各种 XML 时也使用 PULL 解析器，Android 官方推荐开发者们使用 PULL 解析技术。PULL 解析技术是第三方开发的开源技术，它同样可以应用于 JavaSE 开发。

XML 文件由嵌套的标签元素（Tag）组成，标签内含有由键值对组成的属性，在开始标签和结束标签内部可以嵌套子标签或内容是值属性的标签体。PULL 就是通过判断每个节点是开始标签，还是结束标签来获取相对应值的，开发人员使用 XmlPullParser（XML Pull 解析器）类提供的功能函数对 XML 数据进行解析，XmlPullParser 对象由工厂类

XmlPullParserFactory 创建，创建 XmlPullParser 对象的示例代码如下：
```
XmlPullParserFactory factory=XmlPullParserFactory.newInstance();
XmlPullParser xmlPullParser=factory.newPullParser();
```

XmlPullParser 常用方法如下。
- xmlPullParser.getEventType()返回当前节点的事件类型（如 START_TAG, END_TAG, TEXT, etc.)。
- xmlPullParser.getName()获取当前节点对应的名称。
- xmlPullParser.getAttributeCount()获取当前节点对应的属性个数。
- xmlPullParser.getText()获取当前节点对应的文本内容。
- xmlPullParser.getAttributeName()获取属性对应的名称。
- xmlPullParser.getAttributeValue()获取属性对应的值。
- xmlPullParser.next()移动到下一个事件。

【例 9.3】使用 PULL 解析 XML 字符串示例。

（1）创建 XML 文件的代码如下：
```xml
<persons>
    <person id = "1">
        <name>张三</name>
        <age>18</age>
    </person>
    <person id = "2">
        <name>李四</name>
        <age>43</age>
    </person>
</persons>
```

（2）创建 Person 类，对应 XML 文件中的 Java Bean，代码如下：
```java
public class Person {
    private int id;
    private String name;
    private int age;
    public Person() {

    }
    public Person(int id, String name, int age) {
        this.id = id;
        this.name = name;
        this.age = age;
    }
    public int getId() {
        return id;
    }
    public void setId(int id) {
        this.id = id;
    }
    public String getName() {
        return name;
    }
    public void setName(String name) {
        this.name = name;
```

```java
        }
        public int getAge() {
            return age;
        }
        public void setAge(int age) {
            this.age = age;
        }
        @Override
        public String toString() {
            return "姓名:" + this.name + ",年龄:" + this.age;
        }
    }
```

（3）编写使用 PULL 解析 XML 输入流的逻辑代码如下：

```java
        public static ArrayList<Person> getPersons(InputStream xml) throws Exception {
            ArrayList<Person> persons = null;
            Person person = null;
            XmlPullParserFactory factory = XmlPullParserFactory.newInstance();
            XmlPullParser parser = factory.newPullParser();
            parser.setInput(xml, "UTF-8");
            int eventType = parser.getEventType();
            while (eventType != XmlPullParser.END_DOCUMENT) {
                switch (eventType) {
                    case XmlPullParser.START_DOCUMENT:
                        persons = new ArrayList<Person>();
                        break;
                    case XmlPullParser.START_TAG:
                        if ("person".equals(parser.getName())) {
                            person = new Person();
                            int id = Integer.parseInt(
                                    parser.getAttributeValue(0));
                            person.setId(id);
                        } else if ("name".equals(parser.getName())) {
                            String name = parser.nextText();
                            person.setName(name);
                        } else if ("age".equals(parser.getName())) {
                            int age = Integer.parseInt(
                                    parser.nextText());
                            person.setAge(age);
                        }
                        break;
                    case XmlPullParser.END_TAG:
                        if ("person".equals(parser.getName())) {
                            persons.add(person);
                            person = null;
                        }
                        break;
                }
                eventType = parser.next();
            }
            return persons;
        }
```

2．向 Internet 发送 JSON 数据

JSON（JavaScript Object Notation，JS 对象简谱）是一种轻量级的数据交换格式。它基于 ECMAScript（欧洲计算机协会制定的 JS 规范）的一个子集，采用完全独立于编程语言的文本格式来存储和表示数据。简洁和清晰的层次结构使得 JSON 成为理想的数据交换语言。易于人们阅读和编写，同时也易于机器解析和生成，并能有效地提升网络传输效率。JSON 是一种轻量级的数据交换格式，具有良好的可读性，以及快速编写的特性。业内主流技术为其提供了完整的解决方案，从而可以在不同平台间进行数据交换。

JSON 主要用来表示数据对象和数组对象。其中，数据对象由花括号括起来的逗号分割的成员构成，成员是由字符串键和逗号分割的键值对组成，例如：

```
{"name": "John Doe", "age": 18, "address": {"country" : "china", "zip-code": "10000"}}
```

数组对象是由方括号括起来的一组值构成，数组对象的成员可以是数据对象，数据对象内部也可以嵌套数组对象，例如：

```
[
    { "id":"1","name":"张三","age":"18" },
    { "id":"2","name":"李四","age":"18" },
    { "id":"3","name":"王五","age":"18" }
]
```

解析 JSON 数据 Android SDK 提供了完整的支持，这些 API 都存在于 org.json 包下，常用类如下。

- JSONObject：JSON 对象，可以完成 JSON 字符串与 Java 对象的相互转换；
- JSONArray：JSON 数组，可以完成 JSON 字符串与 Java 集合或对象的相互转换；
- JSONStringer：JSON 文本构建类，这个类可以帮助快速创建 JSON text，每个 JSONStringer 实体只能对应创建一个 JSON text；
- JSONTokener：JSON 解析类；
- JSONException：JSON 异常。

解析上面 JSON 字符串的示例代码如下：

```java
private void parseEasyJson(String json){
    persons = new ArrayList<Person>();
    try{
        JSONArray jsonArray = new JSONArray(json);
        for(int i = 0;i < jsonArray.length();i++){
            JSONObject jsonObject = (JSONObject) jsonArray.get(i);
            Person person = new Person();
            person.setId(jsonObject.getString("id"));
            person.setName(jsonObject.getString("name"));
            person.setAge(jsonObject.getString("age"));
            persons.add(person);
        }
    }catch (Exception e){
        e.printStackTrace();
    }
}
```

9.2 Android 网络访问框架

使用网络访问框架可以简化编码，使得开发人员可以将注意力集中在核心逻辑的开发上，较为常用的框架包括 Volley、OKHttp 和 WebView。

9.2.1 Volley

Volley 是一个可让 Android 应用更快捷连接网络的 HTTP 库。Volley 具有如下优点：自动网络请求调度；多个并发网络连接；透明磁盘和具有标准 HTTP 缓存一致性的内存响应缓存；支持请求优先级；取消请求 API，可以取消单个请求，也可以设置要取消的请求时间段或范围；可轻松自定义，如自定义重试和退避时间；强大的排序功能，可以轻松地使用从网络异步提取的数据正确填充界面。

如需使用 Volley，必须向应用清单文件（AndroidManifest.xml）中添加 android.permission.INTERNET 权限。否则，应用将无法连接网络。当然，作为第三方类库，如果要在工程中使用 Volley 必须先引入它，首先执行菜单操作，选择"File"→"Project Structure"菜单命令，在弹出的对话框中选择对应模块（Module），单击"Depencies"按钮，然后单击"+"按钮，选择"Library Dependency"选项，输入"volley"后单击"Search"按钮，选择对应版本库软件后，单击"OK"按钮完成操作，如图 9.6 所示。

图 9.6　添加 Volley

使用 Volley 发起一条 HTTP 请求需要获取一个 RequestQueue 对象，调用如下方法获取：

RequestQueue mQueue = Volley.newRequestQueue(context);

通过 Volley.newRequestQueue 方法可使用默认值设置一个 RequestQueue，并启动该队列。如需发送请求，可以构建一个请求，并使用 add() 将其添加到 RequestQueue 中。调用 add() 时，Volley 会运行一个缓存处理线程和一个网络调度线程池。将请求添加到队列后，缓存线程会获取该请求并对其进行分类，如果该请求可以通过缓存处理，则系统会在缓存线程上解析缓存的响应，并在主线程上传送解析后的响应。如果该请求无法通过缓存处理，则系统会将其放置到网络队列中。第一个可用的网络线程会从队列中获取该请求，执行 HTTP 事务，在工作器线程上解析响应，将响应写入缓存，然后将解析后的响应发送回主线程以供传送。RequestQueue 是一个请求队列对象，它可以先缓存所有的 HTTP 请求，再按照一定的算法并发地发出这些请求。

下面创建一个 StringRequest 对象来发出一条 HTTP 请求，示例代码如下：

```java
StringRequest stringRequest = new StringRequest(
    "https://www.baidu.com",
    new Response.Listener<String>() {
        @Override
        public void onResponse(String response) {
            Log.d("TAG", response);
        }
    }, new Response.ErrorListener() {
        @Override
        public void onErrorResponse(VolleyError error) {
            Log.e("TAG", error.getMessage(), error);
        }
    });
```

使用 new 创建 StringRequest 对象，StringRequest 的构造函数需要传入三个参数，第一个参数是目标服务器的 URL 地址，第二个参数是服务器响应成功的回调，第三个参数是服务器响应失败的回调。最后，将这个 StringRequest 对象添加到 RequestQueue 里就可以使用了，代码如下：

```java
mQueue.add(stringRequest);
```

运行代码，在 LoagCat 选项卡中可以看到如图 9.7 所示的返回结果，网址为 "www.baidu.com" 的 HTML 内容。

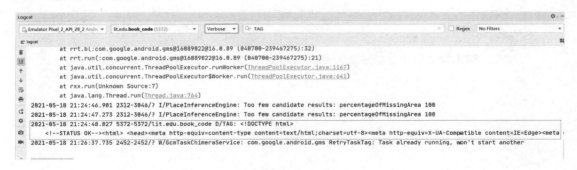

图 9.7 Volley 请求返回结果

发送 POST 请求时，需要在创建 StringRequest 对象中使用另外一种四个参数的构造函数，其中第一个参数是指定请求类型，构造函数声明如下：

```java
StringRequest stringRequest = new StringRequest(Method.POST, url,
    listener, errorListener);
```

但是 Volley 没有提供 POST 请求携带参数的方法，此时需要调用 StringRequest 的父类，即 Request 中的 getParams() 方法来获取 POST 参数，在 StringRequest 的匿名类中重写 getParams() 方法，在这里设置 POST 参数就可以了，重写【例 9.2】的请求代码如下：

```java
String strUrl = "http://192.168.43.33:8080/AndroidWeb/GetPostServlet";
StringRequest stringRequest = new StringRequest(Method.POST, url,
```

```java
            new Response.Listener<String>() {
                @Override
                public void onResponse(String response) {
                    Log.d("TAG", response);
                }
            }, new Response.ErrorListener() {
                @Override
                public void onErrorResponse(VolleyError error) {
                    Log.e("TAG", error.getMessage(), error);
                }
            }) {
        @Override
        protected Map<String, String> getParams() throws
        AuthFailureError {
            Map<String, String> map = new HashMap<String, String>();
            map.put("password", password.getText().toString());
            map.put("userName", userName.getText().toString());
            return map;
        }
    };
```

Volley 库与远程服务器之间可以方便地传输图片、字符串、JSON 对象和 JSON 对象数组，在以 POST 方式发送请求时，如果需要交互大量参数数据，开发人员通常使用 JSON 格式来进行数据封装，Volley 库此时提供 JsonRequest 类加以支持，示例代码如下：

```java
String strUrl = "http://192.168.43.33:8080/AndroidWeb/GetPostServlet";
Map<String, String> map = new HashMap<String, String>();
map.put("password", password.getText().toString());
map.put("userName", userName.getText().toString());
JSONObject jsonObject = new JSONObject(map);
JsonRequest<JSONObject> request = new JsonRequest(Method.POST, url,
        jsonObject, new Response.Listener<String>() {
            @Override
            public void onResponse(String response) {
                Log.d("TAG", response);
            }
        }, new Response.ErrorListener() {
            @Override
            public void onErrorResponse(VolleyError error) {
                Log.e("TAG", error.getMessage(), error);
            }
        }) {
    @Override
    public Map<String, String> getHeaders() {
        HashMap<String, String> headers = new HashMap<String, String>();
        headers.put("Accept", "application/json");
        headers.put("Content-Type",
                "application/json; charset=UTF-8");
        return headers;
    }
};
```

JsonRequest 类构造函数新增的一个参数是 POST 请求传输的 JSON 数据对象，所以此时就没有必要再重写 getParams()方法了，只需重写 getHeaders()方法，在其中设置请求头属性 Accept 的值为 application/json 和属性 Content-Type 的值为 application/json 来保证正常传输 JSON 数据即可。

9.2.2 OkHttp

OkHttp 是一个高效的 HTTP 客户端，它具有以下默认特性：允许所有同一个主机地址的请求共享同一个 Socket 连接；连接池减少请求延时；透明的 GZIP 压缩减少响应数据的大小；缓存响应内容，避免一些完全重复的请求。

在工程中引入 OkHttp 首先执行菜单操作，执行"File"→"Project Structure"菜单命令，在弹出的对话框中选择对应模块（Module）后单击"Depencies"按钮，然后单击"+"按钮，选择"Library Dependency"选项，输入"okhttp"后单击"Search"按钮，选择对应版本库的软件后单击"OK"按钮完成操作，如图 9.8 所示。

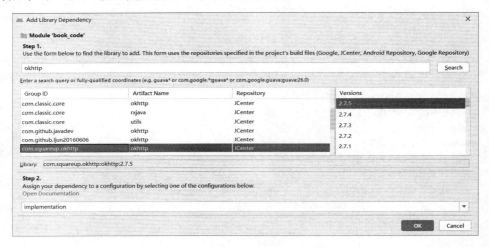

图 9.8　添加 OkHttp

OkHttp 的请求/响应 API 使用构造器模式 builders 来设计，它支持阻塞式的同步请求和带回调的异步请求。

1．GET 请求

创建 OkHttpClient 对象：
```
new OkHttpClient;
```
构造 Request 对象：
```
Request request = new Request.Builder() .url(url) .get() .build();
```
通过前两步中的对象构建 Call 对象，并通过 Call.enqueue(Callback)方法来提交异步请求，示例代码如下：
```
String url = "http://www.baidu.com";
OkHttpClient okHttpClient = new OkHttpClient();
final Request request = new Request.Builder()
        .url(url)
        .get()
        .build();
```

```
            Call call = okHttpClient.newCall(request);
            call.enqueue(new Callback() {
                @Override
                public void onFailure(Request request, IOException e) {
                    Log.d("TAG", "onFailure: ");
                }

                @Override
                public void onResponse(Response response) throws IOException {
                    Log.d("TAG", "onResponse: " + response.body().string());
                }
            });
```

使用 call.enqueue()发送异步请求，同时 OkHttp 也支持同步请求，但是 Android 3.0 后已经不允许在主线程访问网络了，因为网络请求是一个耗时操作有可能会阻塞调用线程，造成 Android App 的 ANR（Application Not Responding）异常，所以此时需要把网络访问相关代码放置在工作线程中，示例代码如下：

```
            String url = "http://wwww.baidu.com";
            OkHttpClient okHttpClient = new OkHttpClient();
            final Request request = new Request.Builder() .url(url) .build();
            final Call call = okHttpClient.newCall(request);
            new Thread(new Runnable() {
                @Override public void run() {
                    try {
                        Response response = call.execute();
                        Log.d(TAG, "run: " + response.body().string());
                    } catch (IOException e) {
                        e.printStackTrace();
                    }
                }
            }).start();
```

无论是同步请求还是异步请求，接收的 Response 对象均在子线程中，其中通过 Response 对象获取请求结果需要在子线程中完成，在得到结果后再切换到 UI 线程改变 UI。

2．POST 请求

使用 POST 方式发送请求与前面的区别就是在构造 Request 对象时，需要多构造一个 RequestBody 对象，用它携带要提交的数据。在构造 RequestBody 时需要指定 MediaType，用于描述请求/响应 body 的内容类型。

RequestBody 是一个抽象类，分别有 FormBody 和 MultipartBody 两个子类。提交表单时，使用 RequestBody 的实现类 FormBody 来描述请求体，它可以携带一些经过编码的 key-value 请求体。示例代码如下：

```
            OkHttpClient client = new OkHttpClient();
            FormBody body = new FormBody.Builder()
                    .add("password", password.getText().toString())
                    .add("userName", userName.getText().toString())
                    .build();
            String url= "http://192.168.43.33:8080/AndroidWeb/GetPostServlet";
            Request request = new Request.Builder()
                    .url(url)
                    .post(body)
```

```
                    .build();
    Call call = client.newCall(request);
    call.enqueue(new Callback() {
            @Override
            public void onFailure(Call call, IOException e) {
                Log.e("TAG", error.getMessage(), error);
            }

            @Override
            public void onResponse(Call call, Response response) throws
                IOException {
                  Log.d("TAG", response);
            }
    });
```

MultipartBody 支持多类型的参数传递,在传输表单类型的参数时还可以传输文件,这时创建一个 MultipartBody 对象再调用 post()方法就可以了。当然,为确保上传的文件程序能够正确运行,还需要在创建 MultipartBody 对象时使用"application/octet-stream"参数,示例代码如下:

```
public void uploadFile(String url, File file) {
    OkHttpClient client = new OkHttpClient();
    RequestBody body =   RequestBody.create(
           MediaType.get("application/octet-stream"), file);
    Request request = new Request.Builder()
            .url(url)
            .post(body)
            .build();
    Call call = client.newCall(request);
    ...
}
```

9.2.3 WebView

WebView 在 Android 平台上是一个特殊的 View,它能用来显示网页,这个 WebView 类可以被用来在 App 中仅显示一个在线的网页,也可以用来开发浏览器。

Android 系统内部包含有内置浏览器,该浏览器内部实现是采用渲染引擎(WebKit)来展示 View 的内容,提供网页前进与后退、网页放大、缩小、搜索等功能。

WebView 是一个基于 WebKit 引擎、展现 Web 页面的控件,Android 的 WebView 在低版本和高版本采用了不同的 WebKit 版本内核。它是 Android 系统中的原生控件,其主要功能是与前端页面进行响应交互,可快捷、省时地实现相关功能,相当于增强版的内置浏览器。WebView 最简单的使用方式是直接显示网页内容,其步骤如下:

(1)在布局文件中添加 WebView 控件;
(2)在代码中让 WebView 控件加载显示网页。

【例 9.4】WebView 示例。

(1)在程序清单文件中添加如下请求权限语句:
```
<uses-permission android:name="android.permission.INTERNET" />
```

(2)编写 XML 布局文件,添加 WebView 控件,代码如下:
```
<WebView
```

```
            android:id="@+id/web_view"
            android:layout_width="match_parent"
            android:layout_height="match_parent">
</WebView>
```

（3）编写 Activity 类文件，代码如下：
```
String url = "https://www.baidu.com";
WebView webView = (WebView) findViewById(R.id.web_view);
webView.loadUrl(url);
```

（4）运行程序结果如图 9.9 所示。

WebView 控件提供了非常丰富的方法为开发这对 WebKit 引擎加以控制，常用方法如下：

- loadUrl()：加载网页。
- loadData()：加载字符串。
- goBack()：执行回退操作。
- reLoad()：刷新页面。

如果想让 WebView 控件支持网页的放大功能和缩小功能，可以通过如下代码进行控制：
```
webview.getSettings().setSupportZoom(true);
webview.getSettings().setBuiltInZoomControls(true);
```

Android 系统还可以使用 WebView 控件嵌入现有的 Web 应用程序，或者是使用 H5+CSS3 来显示丰富的文本效果，这些功能都是借助于 WebView 控件的 loadDataWithBaseURL() 方法，示例代码如下：
```
StringBuilder sb = new StringBuilder();
sb.append("<div>请选择您的爱好：</div>");
sb.append("<ul>");
sb.append("<li>游泳</li>");
sb.append("<li>滑雪</li>");
sb.append("<li>打球</li>");
sb.append("</ul>");
webview.loadDataWithBaseURL(null, sb.toString(),
"text/html", "utf-8", null);
```

运行代码显示结果如图 9.10 所示。

图 9.9　WebView 示例

图 9.10　WebView 显示 HTML

loadDataWithBaseURL(String baseUrl, String data, String mimeType, String encoding, String historyUrl)中参数的含义如下。

- Data：指定需要加载的 HTML 代码。
- mimeType：指定 HTML 代码的 MIME 类型，对于 HTML 代码可指定为 text/html。
- Encoding：指定 HTML 代码所用的字符集，如 UTF-8。

在 HTML 中往往使用 JavaScript 代码进行用户交互、信号处理等功能。默认情况下 WebView 控件不能执行 JavaScript 代码，这时通过 setJavaScriptEnabled()设置改变 WebView 控件的默认功能，代码如下：

```
WebSettings set = webView.getSettings();
set.setJavaScriptEnabled(true);
```

但对于 JavaScript 的 Windows 操作，如弹窗和对话框操作仍然无效，这时就要使用 WebView 控件的 setWebChromeClient()了，代码如下：

```
webView.setWebChromeClient(new WebChromeClient());
```

【例 9.5】WebView 支持 JavaScript 示例。

（1）创建应用程序，添加本地 HTML 资源，选定工程 main 目录，单击鼠标右键，执行"New"→"Folder"→"Assets Folder"菜单命令，在弹出的对话框中单击"Finish"按钮，完成操作，如图 9.11 所示。

图 9.11　创建 Assets 目录

在创建的 Asserts 目录下添加一个 HTML 文件"testJavaScript.html"，文件代码如下：

```html
<html>
<body>
<a><h1>js 中调用本地方法</h1></a>
<script>
    function sayHello(){
        alert("Hello");
    }
    var aTag = document.getElementsByTagName('a')[0];
    aTag.addEventListener('click', function(){
        control.toastMessage("js 中调用本地方法");
```

```
                    return false;
            }, false);
    </script>
    </body>
</html>
```

（2）创建 Activity，命名为"WebViewJavaScriptActivity"，编辑布局 XML 文件，使用纵向线性布局（LinearLayout），添加一个按钮控件，单击可以调用 JavaScript 代码和一个 WebView 控件来显示 HTML 代码，XML 布局文件代码如下：

```xml
<LinearLayout
    xmlns:android="http://schemas.android.com/apk/res/android"
    xmlns:tools="http://schemas.android.com/tools"
    android:layout_width="match_parent"
    android:layout_height="match_parent"
    android:background="#bbe4fb"
    android:orientation="vertical"
    android:paddingLeft="15dp"
    android:paddingTop="15dp"
    android:paddingRight="15dp"
    android:paddingBottom="15dp"
    tools:context=".chap10.s2.WebViewJavaScriptActivity">
    <Button
        android:id="@+id/btn_show"
        android:layout_width="match_parent"
        android:layout_height="wrap_content"
        android:text="本地调用 JS 方法" />
    <WebView
        android:id="@+id/webView"
        android:layout_width="match_parent"
        android:layout_height="match_parent"></WebView>
</LinearLayout>
```

（3）编写 Activity 代码，在 onCreate 函数内部添加属性初始化逻辑，代码如下：

```java
@Override
protected void onCreate(Bundle savedInstanceState) {
    super.onCreate(savedInstanceState);
    setContentView(R.layout.activity_web_view_java_script);
    button = (Button) findViewById(R.id.btn);
    webView = (WebView) findViewById(R.id.webView);
    webView.setWebChromeClient(new WebChromeClient());
    WebSettings set = webView.getSettings();
    set.setJavaScriptEnabled(true);
    webView.addJavascriptInterface(new JsInteraction(), "control");
    webView.loadUrl("file://android_asset/testJavaScript.html");
    button.setOnClickListener(new View.OnClickListener() {
        @Override
        public void onClick(View view) {
            webView.loadUrl("javascript:sayHello()");
        }
    });
}
```

在代码中为按钮控件设置单击"事件监听器"按钮，并在事件处理函数后直接运行

JavaScript 代码，显示 alert 窗口，如图 9.12（a）所示。而单击 WebView 控件内的超链接时并没有显示 alert 窗口，而显示了 Android 系统的 Toast 窗口，显示结果如图 9.12（b）所示，这是因为在代码中为 WebView 控件设置了处理接口，代码如下：

```
public class JsInteraction {
    @JavascriptInterface
    public void toastMessage(String message) {
        Toast.makeText(getApplicationContext(), message,
            Toast.LENGTH_LONG).show();
    }
}
```

程序中通过"webView.addJavascriptInterface(new JsInteraction(), "control");"为控件添加 JavaScript 处理接口，其中，函数的第一个参数为接口对象，第二个参数为接口的名称。

（a）本地调用 JS 代码　　　　　　　　（b）HTML 调用 JS 代码

图 9.12　在 WebView 控件内调用 JavaScript 代码

9.3　Socket 网络编程

HTTP 属于应用层，Android 网络编程不仅可以基于 HTTP，还可以基于下层网络的通信协议，如基于传输层的 TCP/IP 和 UDP 进行网络应用程序开发，这时就需要 Android SDK 对 Socket 的支持了。

Socket（套接字）是应用层与 TCP/IP 协议族通信的中间软件抽象层，Socket 属于传输层，因为 TCP / IP 属于传输层，所以 Socket 解决的是数据如何在网络中传输的问题，而 HTTP 属于应用层，解决的是如何包装数据。Socket 是一组接口，在设计模式中 Socket 就是一个门面模式，它把复杂的 TCP/IP 协议族隐藏在 Socket 接口后面，对用户来说使用 Socket 接口来连接网络，让 Socket 去组织数据，以符合指定的协议完成应用程序网络数据传输的要求。Socket 在网络编程中所处位置如图 9.13 所示。

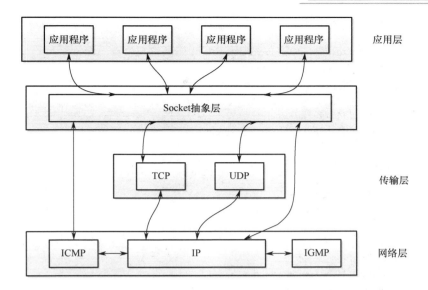

图9.13　Socket

Socket 是 "open—write/read—close" 模式的一种实现。在网络编程时，服务器端先初始化 Socket，然后与端口绑定（bind），对端口进行监听（listen），调用 accept 函数阻塞进程等待客户端连接。在这时如果客户端初始化了一个 Socket，然后连接服务器，如果连接成功，客户端与服务器端的连接就建立了。客户端先发送数据请求，服务器端接收请求并处理请求，然后把回应数据发送给客户端，客户端读取数据后关闭连接，即一次交互结束，运行流程如图 9.14 所示。

Socket 的使用类型主要有两种：一种是流套接字（Streamsocket），基于 TCP 采用流的方式提供可靠的字节流服务；另一种是数据报套接字（Datagramsocket），基于 UDP 采用数据报文提供数据打包发送的服务。下面以 TCP 为例讲解 Android Socket 编程方法。

【例 9.6】Android Socket 编程示例。

（1）创建 Android 应用程序作为 Socket 客户端，编辑 XML 布局文件，代码如下：

```
<?xml version="1.0" encoding="utf-8"?>
<LinearLayout
xmlns:android="http://schemas.android.com/apk/res/android"
    xmlns:app="http://schemas.android.com/apk/res-auto"
    xmlns:tools="http://schemas.android.com/tools"
    android:layout_width="match_parent"
    android:layout_height="match_parent"
    android:orientation="vertical"
    tools:context=".chap10.s2.SocketActivity">
    <Button
        android:id="@+id/connect"
        android:layout_width="match_parent"
        android:layout_height="wrap_content"
        android:text="连接" />
    <Button
        android:id="@+id/disconnect"
        android:layout_width="match_parent"
        android:layout_height="wrap_content"
```

```xml
            android:text="断开" />
        <TextView
            android:id="@+id/receive_message"
            android:layout_width="match_parent"
            android:layout_height="wrap_content" />
        <Button
            android:id="@+id/Receive"
            android:layout_width="match_parent"
            android:layout_height="wrap_content"
            android:text="接收" />
        <EditText
            android:id="@+id/edit"
            android:layout_width="match_parent"
            android:layout_height="wrap_content" />
        <Button
            android:id="@+id/send"
            android:layout_width="match_parent"
            android:layout_height="wrap_content"
            android:text="发送" />
</LinearLayout>
```

程序采用纵向线性布局（LinearLayout），在其内部添加四个按钮，通过单击分别执行连接、发送、接收和断开连接的操作，另外，还有一个静态文本控件（TextView）来显示服务器发送的数据，以及一个文本编辑控件（EditText）来编辑需要发送给服务器的文本，运行程序如图 9.15 所示。

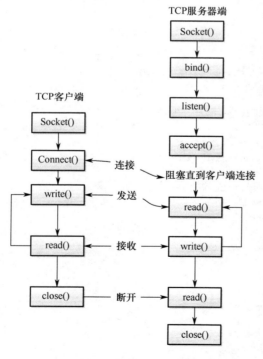

图 9.14　Socket 流程　　　　　图 9.15　Socket 示例界面

（2）编辑 Activity 代码，添加控件属性如下：

```java
private Button btnConnect, btnDisconnect, btnSend, receive;
private TextView receive_message;
private EditText mEdit;
```

在 onCreate() 内分别进行初始化，代码如下：

```java
btnConnect = (Button) findViewById(R.id.connect);
btnDisconnect = (Button) findViewById(R.id.disconnect);
btnSend = (Button) findViewById(R.id.send);
receive = (Button) findViewById(R.id.Receive);
mEdit = (EditText) findViewById(R.id.edit);
receive_message = (TextView) findViewById(R.id.receive_message);
```

修改 Activity 类的声明，使其实现 View.OnClickListener 接口，同时也是按钮控件的单击事件处理器，这时为按钮对象添加时间处理器，代码如下：

```java
btnConnect.setOnClickListener(this);
btnDisconnect.setOnClickListener(this);
btnSend.setOnClickListener(this);
receive.setOnClickListener(this);
```

添加和 Socket 通信相关的属性，代码如下：

```java
private Socket socket;
private ExecutorService mThreadPool;
InputStream is;
InputStreamReader isr;
BufferedReader br;
String response;
OutputStream outputStream;
```

需要说明的是，在 Android 主线程中不能进行网络操作，所以本示例使用了 ExecutorService 类，并在 onCreate() 内完成初始化。

程序需要重写按钮（单击事件处理）函数，代码如下：

```java
@Override
public void onClick(View view) {
    switch (view.getId()) {
        case R.id.connect:
            mThreadPool.execute(new Runnable() {
                @Override
                public void run() {
                    try {
                        socket = new Socket("192.168.43.33", 8099);
                    } catch (IOException e) {
                        e.printStackTrace();
                    }
                }
            });
        case R.id.receive_message:
            mThreadPool.execute(new Runnable() {
                @Override
                public void run() {
                    try {
                        is = socket.getInputStream();
                        isr = new InputStreamReader(is);
                        br = new BufferedReader(isr);
```

```java
                            response = br.readLine();
                            Message msg = Message.obtain();
                            msg.what = 0;
                            mMainHandler.sendMessage(msg);
                        } catch (IOException e) {
                            e.printStackTrace();
                        }
                    }
                });
            case R.id.send:
                mThreadPool.execute(new Runnable() {
                    @Override
                    public void run() {
                        try {
                            outputStream = socket.getOutputStream();
                            outputStream.write((mEdit.getText()
                                .toString() + "\n").getBytes("utf-8"));
                            outputStream.flush();
                        } catch (IOException e) {
                            e.printStackTrace();
                        }
                    }
                });
            case R.id.disconnect:
                try {
                    outputStream.close();
                    br.close();
                    socket.close();
                } catch (IOException e) {
                    e.printStackTrace();
                }
        }
```

代码中先通过 ID 来判断用户单击了哪个按钮，再分别执行不同的代码，需要注意的是，所有的网络操作都需要在工作线程内完成。从代码中可以看出客户端从服务器获取数据，以及向服务器发送数据所执行的操作都是通常的流操作，这说明 Socket 完全隐藏了 TCP/IP 的底层操作。当然，客户端从服务器得到数据后是需要显示在程序界面上的，因为这些代码运行在工作线程中，而 Android 不允许在工作线程中改变程序控件的状态，这时候就需要使用 Android 的 Handler 机制来处理消息了，代码如下：

```java
private Handler mMainHandler = new Handler() {
    @Override
    public void handleMessage(Message msg) {
        switch (msg.what) {
            case 0:
                receive_message.setText(response);
                break;
        }
    }
};
```

（3）创建 Socket 服务器。为简化设计过程本示例采用了 Apache 的 Mina 框架，编写网络事件处理代码如下：

```java
public class SocketHandler extends IoHandlerAdapter {
    @Override
    public void exceptionCaught(IoSession session, Throwable cause) throws Exception {
        System.out.println("exceptionCaught: " + cause);
    }
    @Override
    public void messageReceived(IoSession session, Object message) throws Exception {
        System.out.println("recieve : " + (String) message);
        session.write("Server responce");
    }
    @Override
    public void messageSent(IoSession session, Object message) throws Exception {
    }
    @Override
    public void sessionClosed(IoSession session) throws Exception {
        System.out.println("sessionClosed");
    }
    @Override
    public void sessionOpened(IoSession session) throws Exception {
        System.out.println("sessionOpen");
    }
    @Override
    public void sessionIdle(IoSession session, IdleStatus status) throws Exception {
    }
}
```

在 Main 框架下，不同的网络事件发生时分别调用不同的代码，当然为了建设服务器还需要一个独立应用，代码如下：

```java
public class SocketServer {
    public static void main(String[] args) {
        NioSocketAcceptor acceptor = null;
        try {
            acceptor = new NioSocketAcceptor();
            acceptor.setHandler((IoHandler) new SocketHandler());
            acceptor.getFilterChain().addLast("mFilter", new
                    ProtocolCodecFilter(new TextLineCodecFactory()));
            acceptor.setReuseAddress(true);
            acceptor.bind(new InetSocketAddress(8099));
        } catch (Exception e) {
            e.printStackTrace();
        }
    }
}
```

（4）运行程序。

运行 Socket 服务器，启动 Android 客户端应用程序，单击"连接"按钮，在服务器端 console 窗口内可以看到连接建立提示，如图 9.16 所示。

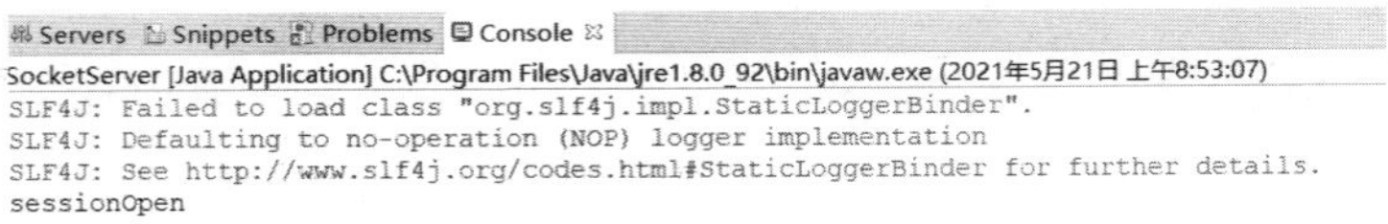

图 9.16　Socket 建立连接

在 Android 客户端输出编辑框中输入字符串，单击"发送"按钮，操作如图 9.17 所示。此时可以在服务器端看到客户端所发送的数据，如图 9.18 所示。

图 9.17　Socket 客户端发送数据　　　　图 9.18　Socket 服务器端接收数据

单击"接收"按钮，客户端显示了服务器端发送过来的数据，显示结果如图 9.19 所示。

单击"断开"按钮，服务器端在 console 窗口中显示了断开连接的提示，显示结果如图 9.20 所示。

图 9.19　Socket 客户端接收数据　　　　图 9.20　断开连接

习题 9

1．设计一个新闻浏览程序，具体要求如下。

（1）架设 Web 服务器，新闻内容通过网络获取，格式为 JSON。将新闻数据文件保存在服务器上，文件内容和格式示例如下。

```
[
    {
        "title":"新冠肺炎疫情速递",
        "content":"××省无新增确诊病例"
    },
    {
        "title":"天气预报",
        "content":"五一未过，沙尘又起"
    },
    {
        "title":"南北车正式公布合并",
        "content":"南北车将于今日正式公布合并"
    },
    {
        "title":"史海钩沉",
        "content":"1955 年授衔将帅录"
    },
    {
        "title":"最美公路",
        "content":"最美公路  难以想象"
    }
]
```

（2）编写新闻浏览程序，开启后程序连接新闻服务器，先下载新闻数据 JSON 文件，再解析新闻数据，通过 ListView 控件显示出来。

（3）网络连接分别使用 HttpURLConnector、Volley、OkHttp 等技术，并比较其优/缺点。

2．编写一个网络浏览器软件，在界面输入框中输入网址，单击"确定"按钮，使用 WebView 控件显示网页。

第 10 章　社区服务系统

通过前面各章的学习和实践，相信大家已经掌握了 Android 应用程序开发所使用的基本知识和开发方法。本章将带领大家从一个实际的案例出发，综合运用布局、Activity、Service、Notification 等各种知识，结合 Web 服务器端开发，实现一个功能较为完整、结构较为复杂的综合性应用程序。可促进对前面所学知识的深入理解和掌握，提高编程实践中解决问题的能力。

10.1　项目简介

社区服务系统主要实现社区住户生活用品采购及日常维修服务等功能。社区住户可以通过本程序提交物品购买需求和家政维修需求，社区服务人员可以在线接单并提供相应服务，社区住户也可以对社区服务进行评价。

本系统有两类用户，即社区住户和社区服务人员，需提供针对这两类用户的注册、登录、修改密码等功能。

社区住户在登录后可以提交物品购买或家政维修等需求；查看近期的服务记录，包括两周内已完成及正在进行的服务的状态（未接单、已接单、已完成、已评价）；也可以查看全部的服务记录。

社区服务人员在登录后可以查看社区住户的物品购买需求；查看社区住户家政维修需求；接受以上需求订单并提供服务；查看近期自己服务的订单状态；查看近期（近两周）的服务记录。

10.2　功能需求

社区服务系统分为 Android 手机端和 Web 服务器端。Android 手机端提供与用户的交互界面，并实现与服务器端的数据传递。Web 服务器端接收用户请求，操作数据库，并返回结果给 Android 手机端。

10.2.1　Android 手机端

根据系统功能需求，将 Android 手机端的功能划分为以下模块。

1. 登录注册功能模块

社区的住户和服务人员的注册信息均存储于后台数据库中。用户注册时系统将数据提交给 Web 服务器端并保存于数据库。

2．社区住户功能模块

（1）发布需求

社区住户发布自己的购物或家政维修请求，并等待服务人员接单。

（2）查看订单

社区住户可以随时查看当前订单的状态。在社区服务人员接单后，当前订单改为已接单，并显示服务人员的姓名、电话等信息。服务人员开始送货或开始维修服务时，订单状态改为服务中。服务人员送货完毕，或维修服务完成，当面付款后，订单状态为已完成（由服务人员完成）。

（3）评价

住户可对已完成的服务订单进行评价。

以上订单信息及服务信息均需存储于 Web 服务器端中。

3．社区服务人员功能模块

（1）查看订单

社区服务人员可以查看自己负责的小区内的所有服务订单。

（2）接单、修改订单状态

社区服务人员可根据自身时间安排及业务能力，接收可以服务的订单。接单后，当前订单改为已接单，在订单界面显示住户的姓名、电话、具体地址、购物或维修需求的详细信息。服务人员开始送货或开始维修服务，订单状态改为服务中。服务人员送货完毕，或维修服务完成，当面付款后，订单状态为已完成。

以上订单信息及服务信息均需存储于 Web 服务器端中。

Android 手机端功能结构如图 10.1 所示。

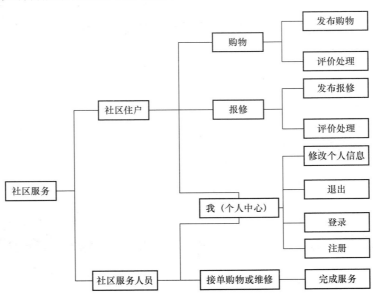

图 10.1　Android 手机端功能结构

10.2.2 Web 服务器端

根据 Android 手机端的请求对需要的数据进行查询、增加、删除、更改等操作，并向用户返回需要的操作结果。因为本程序只需要使用手机进行操作，所以在 Web 服务器端上没有设计操作界面。

10.3 效果展示

社区住户和社区服务人员的界面是不同的，社区住户登录后，跳转到首页，首页底部导航栏有"主页"、"购物"、"维修"和"我"四个选项卡，社区服务人员登录后跳转到订单处理界面，底部导航栏只有"购物"、"维修"和"我"三个选项卡。

社区住户主要界面如图 10.2～图 10.10 所示。

图 10.2　登录

图 10.3　注册

图 10.4　未登录时的首页

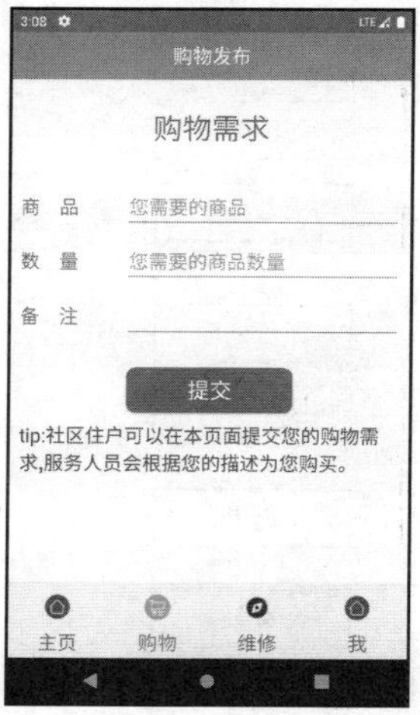

图 10.5　发布购物需求

图 10.6　发布维修需求

图 10.7　我（个人中心）

图10.8　查看购物记录　　图10.9　查看报修记录　　图10.10　修改个人信息

社区服务人员的界面除了与住户相同的登录、注册界面和个人中心界面，还有如图10.11～10.14所示界面。

图10.11　未处理购物订单　　图10.12　未处理维修订单

图10.13　新订单提醒　　图10.14　超时提醒

10.4 系统设计与实现

10.4.1 数据库设计

本项目的数据存放在后台 MySQL 数据库中，根据系统功能需求，包括如下四个数据表，如表 10.1～表 10.4 所示。

1. 用户表 user

表 10.1　user 表

序　号	字 段 名	类　型	主键/外键	中 文 含 义	备 注 说 明
1	Id	Int	是	住户 ID	自增
2	username	Varchar		用户真实姓名	
3	nicname	Varchar		用户昵称	
4	pwd	Varchar		密码	
5	phone	Varchar		联系电话	登录账号
6	address	Varchar		住址	
7	number	Varchar		身份证号	
8	pic	Varchar		头像 URL	

2. 服务人员表 serviceuser

表 10.2　serviceuser 表

序　号	字 段 名	类　型	主键/外键	中 文 含 义	备 注 说 明
1	Id	Int	是	住户 ID	自增
2	username	Varchar		用户真实姓名	
3	nicname	Varchar		用户昵称	
4	pwd	Varchar		密码	
5	phone	Varchar		联系电话	登录账号
6	number	Varchar		身份证号	
7	pic	Varchar		头像 URL	

3. 维修表 repair

表 10.3　repair 表

序　号	字 段 名	类　型	主键/外键	中 文 含 义	备 注 说 明
1	Id	Int	是	报修 ID	自增
2	title	Varchar		标题	
3	detail	Varchar		内容	
4	userid	Int		下单住户 ID	
5	username	Varchar		下单住户昵称	
6	userphone	Varchar		用户手机号	
7	serviceid	Int		服务人员 ID	

(续表)

序 号	字 段 名	类 型	主键/外键	中文含义	备 注 说 明
8	servicename	Varchar		服务人员昵称	
9	status	Varchar		服务状态	
10	level	Int		评价等级	
11	describe	Varchar		评价内容	
12	createtime	timestamp		创建时间	
13	recetime	timestamp		接单时间	
14	starttime	timestamp		开始服务时间	
15	overtime	timestamp		完成时间	
16	evatime	timestamp		评价时间	

4．购物表 purchase 表

表 10.4　purchase 表

序 号	字 段 名	类 型	主键/外键	中文含义	备 注 说 明
1	Id	Int	是	购物 ID	自增
2	coname	Varchar		名称	
3	num	Int		数量	
4	userid	Int		下单住户 ID	
5	username	Varchar		下单住户昵称	
6	userphone	Varchar		用户手机号	
7	serviceid	Int		服务人员 ID	
8	servicename	Varchar		服务人员昵称	
9	status	Varchar		服务状态	
10	level	Int		评价等级	
11	describe	Varchar		评价内容	
12	createtime	timestamp		创建时间	
13	recetime	timestamp		接单时间	
14	starttime	timestamp		开始服务时间	
15	overtime	timestamp		完成时间	
16	evatime	timestamp		评价时间	

10.4.2　Web 服务器端设计

Web 服务器端采用 Springboot+Mybatis 框架，由实体层 entity、工具类接口层 dao、服务层 service 和控制层 controller 组成。

entity 层确定了各个实体的属性，它和 Android 中的实体类对应。dao 接口层规定了对每个数据表的操作，体现为抽象方法，是数据表操作的最底层。service 层包含接口和对应实现类，接口确定了这个服务层包含的功能，如具体查询、增加等；实现类调用 dao 接口实现对应 service 接口的方法。controller 层通过@RestController 和@RequestMapping 两个注解定义服务器对外的访问接口，对程序进行控制。

Web 服务器端主要功能如下。

（1）用户控制。

提供查询用户信息、新增用户（注册）、修改用户信息等功能。

（2）购物控制。

实现对购物订单的操作，将社区住户提交的购物订单信息保存到数据库中。社区服务人员接单后该订单的状态信息会根据不同的服务阶段而改变。社区住户除了可对订单进行评价和描述，还可以查询所有的购物订单信息。

（3）维修控制。

提供对维修订单的操作，社区住户新建维修需求体现为向数据库对应表中添加记录，将社区服务人员接单和更改订单状态视为对应数据项的修改和更新，社区住户评价时可将对应评价星级和描述增加至对应数据项。此外还提供对所有维修订单的查询筛选功能，可以返回对应社区住户 ID 或社区服务人员 ID 的维修订单列表。

下面介绍 Web 服务器端各个功能的实现过程。

1．用户控制功能的实现

（1）在实体层 entity 中创建社区住户和社区服务人员的实体类 User 和 ServiceUser，确定实体的属性内容并在数据库中创建对应的表。使用@Data 注解为实体类自动配置相对应的 Get 和 Set，以及 Tostring 方法。

（2）在 dao 层中创建 UserDao 和 ServiceUserDao，并根据需求定义对数据表的操作，体现为抽象方法。使用@Mapper 将该层标记为 mapper，并在配置文件 application.yml 中使用 mapper-locations: classpath:mapper/*Mapper.xml 实现同对应 mapper.xml 文件的连接。

（3）在 resources 下创建 mapper 文件夹，并在 mapper 目录中创建 usermapper.xml 文件和 ServiceUsermapper.xml 文件，写入对应功能的 SQL 语句，实现基本的增、删、改、查功能。mapper.xml 作为配置文件是操纵数据库的直接方法，使用 namespace 确定该文件与其连接的对象，ID 直接与 dao 层中定义的抽象方法名连接，以确定对应的抽象方法，resultType 确定该方法的返回值类型，parameterType 确定该方法所需要的参数类型。

（4）创建 service 层，service 层包含接口类和实现类。@Service 注解标记该实现类成为服务层，接口类确定了这个服务层包含的具体功能，实现类则是在继承了对应接口类后，使用 dao 层函数对象完成对应的功能对数据库的操作。

（5）创建 controller 层，在 service 层实现相关功能对数据库的操作之后，通过控制层根据外部的需求使用@RestController 和@RequestMapping 两个注解定义服务器对外的访问接口（findpur 接口）对所有用户信息的查询，返回用户列表；findUserByPhone 接口实现登录用户功能，返回用户信息；findUserById 接口实现根据 ID 查询用户信息，返回用户信息；updateUserInfo 接口实现修改用户信息的功能；addUser 接口接收传递的用户实体以实现用户注册功能；updaUserPwd 接口根据 ID 和新密码自动创建 user 实体实现用户修改密码功能；deluser 接口依据 ID 对用户实现删除功能。

其核心代码如下所示。

（1）社区住户 entity 层示例代码：

```
@Data
```

```java
public class User {
    private int id;              //住户 ID
    private String username;     //用户真实姓名
    private String nickname;     //用户昵称
    private String pwd;          //密码
    private String phone;        //联系电话
    private String address;      //住址
    private String number;       //身份证号码
    private String pic;          //头像 URL
}
```

（2）社区住户 dao 层示例代码：

```java
@Mapper
@Repository
public interface UserDao {
    List<User> list();
    User login(String phone);
    User qryByUserId(Integer id);
    int insert(User user);
    int update(User user);
    int updPwd(User user);
    int del(Integer id);
}
```

（3）社区住户 mapper.xml 示例代码：

```xml
<mapper namespace="communityservice.demo.dao.UserDao">
    <select id="list" resultType="communityservice.demo.entity.User">
        SELECT * FROM user
    </select>
    <select id="login" parameterType="java.lang.String" resultType="communityservice.demo.entity.User">
        select *
        from user
        where phone = #{phone}
    </select>
</mapper>
```

（4）社区住户 service 层示例代码：

```java
@Service
public class UserServiceImpl implements UserService {
    @Autowired
    private UserDao userDao;
    @Override
    public List<User> list() {return userDao.list(); }
    @Override
    public User login(String phone) {return userDao.login(phone);}
    @Override
    public User qryByUserId(Integer id) {return userDao.qryByUserId(id); }
    @Override
    public int insert(User user) {return userDao.insert(user);    //添加}
    @Override
    public int update(User user) {return userDao.update(user);    //更新 }
    @Override
    public int updPwd(User user) { return userDao.updPwd(user);}
    @Override
```

```
        public int del(Integer id) {return userDao.del(id);}
    }
```

（5）社区住户 controller 层示例代码：
```
@RestController
public class UserController {
    @Autowired
    private UserService userService;
    @GetMapping("/findAllUser")    //查询所有用户
    public List<User> userlist(){return userService.list(); }
    @RequestMapping("/findUserByPhone")    //登录用户
    public User loginuser(@RequestParam String phone,@RequestParam String pwd) {
        User user = userService.login(phone);
        if (!Objects.isNull(user)) {
            if (Objects.equals(pwd, user.getPwd())) { return user;}
            else { return null;    //0 表示没有用户}
        }else{ return null;}
    }
//修改密码
@RequestMapping("/updateUserPwd")
public boolean updPwdSub(@RequestParam Integer id,@RequestParam String pwd){
    User user=userService.qryByUserId(id);
    user.setPwd(pwd);
    try{
        userService.updPwd(user);
        return true;
    }catch (Exception e){
        e.printStackTrace();
        return false;
    }
}
}
```

2．购物控制功能的实现

（1）在实体层 entity 中创建 Purchase 实体类，并定义购物的属性，使用@Data 注解为实体属性连接数据库，以及在数据库中创建对应的表。

（2）在 dao 层中创建 PurchaseDao，并根据需求定义对应的方法。

（3）在 resources 下创建 mapper 文件夹，并在 mapper 目录下创建 PurchaseMapperx.xml 文件，写入对应功能的 SQL 语句，实现基本的增、删、改、查功能。

（4）创建 service 调用 dao 层接口，包含这个服务层要实现的功能，@Service 注解标记该实现类成为服务层。

（5）通过 controller 层实现 findpur 接口对所有购物订单的查询，返回订单列表；使用 addpur 接口进行订单添加；通过 updatepurstatus 接口更新订单状态信息，并通过调用 setpurRecetime、setpurStarttime、setpurOvertime、setpurEvatime 接口来获取不同服务状态下的当前时间。

购物模块 dao 层主要代码如下：
```
public interface PurchaseDao {
    //1.查询已购商品信息
    public List<Purchase> findAllPurchase();
    //2.购买商品
```

```java
        public void addPurchase(Purchase purchase);
        //3.更改购买商品
        public void updatePurchase(Purchase purchase);
        //4.删除购买商品
        public void delPurchase(Integer id);
        //5.根据服务状态查询商品信息
        public List<Purchase> findByStatus(String status);
        //6.更改购买商品服务状态信息
        public void updatePurchaseStatus(Purchase purchase);
        //7.设置时间
        public void setPurchaseRecetime(Purchase purchase );
         public void setPurchaseStarttime(Purchase purchase);
         public void setPurchaseOvertime(Purchase purchase);
         public void setPurchaseEvatime(Purchase purchase);
}
```

购物模块 service 层主要代码如下：

```java
        public class PurchaseImpl implements PurchaseService {
            @Autowired
            private PurchaseDao purchaseDao;
            @Override
            public List<Purchase> findAllPurchase() {
                List<Purchase> allPurchase = purchaseDao.findAllPurchase();
                return allPurchase;
            }
            @Override
            public void addPurchase(Purchase purchase) {
                purchaseDao.addPurchase(purchase);
            }
            @Override
            public void updatePurchase(Purchase purchase) {
                purchaseDao.updatePurchase(purchase);
            }
            @Override
            public void delPurchase(Integer id) {
                purchaseDao.delPurchase(id);
            }
            @Override
            public List<Purchase> findByStatus(String status) {
                List<Purchase> byStatus = purchaseDao.findByStatus(status);
                return byStatus;
            }
            @Override
            public void updatePurchaseStatus(Purchase purchase) {
                purchaseDao.updatePurchaseStatus(purchase);
            }
            @Override
            public void setPurchaseRecetime(Purchase purchase) {
                purchaseDao.setPurchaseRecetime(purchase);
            }
            @Override
            public void setPurchaseStarttime(Purchase purchase) {
                purchaseDao.setPurchaseStarttime(purchase);
            }
```

```java
        @Override
        public void setPurchaseOvertime(Purchase purchase) {
            purchaseDao.setPurchaseOvertime(purchase);
        }
        @Override
        public void setPurchaseEvatime(Purchase purchase) {
            purchaseDao.setPurchaseEvatime(purchase);
        }
    }
```

购物模块 controller 层主要代码如下:

```java
    public class PurchaseController {
        @Autowired
        private PurchaseService purchaseService;
        String strDateFormat = "yyyy-MM-dd HH:mm:ss";
        SimpleDateFormat sdf = new SimpleDateFormat(strDateFormat);
        @RequestMapping("/findpur")
        public List<Purchase> findAlpurchaseService(){
            List<Purchase>allPurchase=purchaseService.findAllPurchase();
            return allPurchase;
        }
        @RequestMapping("/addpur")
        public void addPurchase(Purchase purchase){
            purchaseService.addPurchase(purchase);
        }
        @RequestMapping("/updatepur")
        public void updatePurchase(Purchase purchase){
            purchaseService.updatePurchase(purchase);
        }
    @RequestMapping("/updatepurstatus")
        public boolean updatePurchaseStatus(Purchase purchase){
            try{
                purchaseService.updatePurchaseStatus(purchase);
                switch (purchase.getStatus()){
                    case "已接单":
                        this.setPurchaseRecetime(purchase);
                        break;
                    case "服务中":
                        this.setPurchaseStarttime(purchase);
                        break;
                    case "已完成":
                        this.setPurchaseOvertime(purchase);
                        break;
                    case "已评价":
                        this.setPurchaseEvatime(purchase);
                        break;
                    default:
                        System.out.println("状态不对");
                        break;
                }
                return true;
            }catch (Exception e){
                e.printStackTrace();
                return false;
```

```
            }
        }
        @RequestMapping("/setpurRecetime")
        Publicboolean    setPurchaseRecetime(@RequestBody    Purchase purchase){
        try{
            System.out.println("setRepairRecetime");
            purchase.setRecetime(sdf.parse(sdf.format(new Date())));
            purchaseService.setPurchaseRecetime(purchase);
            System.out.println("recetime:"+sdf.parse(sdf.format(new Date())));
            return true;
            }catch (Exception e){
                e.printStackTrace();
                return false;
            }
        }
    }
```

3．维修控制功能的实现

维修与购物实现方法类似，在此不再详细介绍，有兴趣的读者可参考相关代码。

10.4.3　Android 手机端的设计与实现

Android 手机端的项目结构如图 10.15 所示。

图 10.15　Android 手机端的项目结构

下面介绍 Android 手机端系统主要功能的设计思路和实现过程。

1. 首页

软件运行后直接进入首页（见图 10.4）。在首页上部显示小区新闻和新鲜事，底部导航栏实现不同界面的切换。

(1) 搭建标题栏布局。

由于大部分界面都有一个标题栏，为了便于代码重复使用，将标题栏抽取出来单独放在一个布局文件 main_title_bar.xml 中，其代码如下：

```xml
<?xml version="1.0" encoding="utf-8"?>
<RelativeLayout xmlns:android="http://schemas.android.com/apk/res/android"
    android:id="@+id/title_bar"
    android:layout_width="match_parent"
    android:layout_height="50dp"
    android:background="@android:color/transparent">
    <TextView
        android:id="@+id/tv_back"
        android:layout_width="50dp"
        android:layout_height="50dp"
        android:layout_alignParentLeft="true"
        android:layout_centerVertical="true"
        android:background="@drawable/go_back_selector"/>
    <TextView
        android:id="@+id/tv_main_title"
        android:layout_width="wrap_content"
        android:layout_height="wrap_content"
        android:textColor="@android:color/white"
        android:textSize="20sp"
        android:layout_centerInParent="true"/>
</RelativeLayout>
```

(2) 首页框架设计。

首页框架在多个地方用到，首页整体为 RelativeLayout 布局，内嵌两个 LinearLayout 布局。第一个 LinearLayout 布局方式为线性垂直，上部嵌入标题栏布局，下部为一个 FrameLayout 布局，用来放置不同的自定义 View 视图。第二个 LinearLayout 布局方式为线性水平，内嵌四个 RelativeLayout 布局，分别为导航栏的四个切换选项卡，每个选项卡都由一个 TextView 控件和一个 ImageView 控件组成。

首页布局代码如下：

```xml
<?xml version="1.0" encoding="utf-8"?>
<RelativeLayout xmlns:android="http://schemas.android.com/apk/res/android"
    xmlns:app="http://schemas.android.com/apk/res-auto"
    android:layout_width="match_parent"
    android:layout_height="wrap_content"
    android:orientation="vertical">
    <LinearLayout
        android:layout_width="match_parent"
        android:layout_height="match_parent"
        android:background="@android:color/white"
        android:orientation="vertical">
        <include layout="@layout/main_title_bar"/>
        <FrameLayout
            android:id="@+id/main_body"
```

```xml
            android:layout_width="match_parent"
            android:layout_height="match_parent"
            android:background="@android:color/white" />
    </LinearLayout>
    <LinearLayout
        android:id="@+id/main_bottom_bar"
        android:layout_width="match_parent"
        android:layout_height="80dp"
        android:layout_alignParentBottom="true"
        android:background="#F2F2F2"
        android:orientation="horizontal">
        <RelativeLayout
            android:id="@+id/bottom_bar_home_btn"
            android:layout_width="0dp"
            android:layout_height="match_parent"
            android:layout_weight="1">
            <TextView
                android:id="@+id/bottom_bar_text_home"
                android:layout_width="match_parent"
                android:layout_height="wrap_content"
                android:layout_alignParentBottom="true"
                android:layout_centerHorizontal="true"
                android:layout_marginBottom="3dp"
                android:gravity="center"
                android:singleLine="true"
                android:text="主页"
                android:textColor="#666666"
                android:textSize="20sp" />
            <ImageView
                android:id="@+id/bottom_bar_image_home"
                android:layout_width="27dp"
                android:layout_height="27dp"
                android:layout_above="@+id/bottom_bar_text_home"
                android:layout_alignParentTop="true"
                android:layout_centerHorizontal="true"
                android:layout_marginTop="3dp"
                android:src="@drawable/icon_firstpage" />
        </RelativeLayout>
        <RelativeLayout
            android:id="@+id/bottom_bar_gouwu_btn"
            android:layout_width="0dp"
            android:layout_height="match_parent"
            android:layout_weight="1">
            <TextView
                android:id="@+id/bottom_bar_text_gouwu"
                android:layout_width="match_parent"
                android:layout_height="wrap_content"
                android:layout_alignParentBottom="true"
                android:layout_centerHorizontal="true"
                android:layout_marginBottom="3dp"
                android:gravity="center"
                android:singleLine="true"
                android:text="购物"
```

```xml
            android:textColor="#666666"
            android:textSize="20sp" />
        <ImageView
            android:id="@+id/bottom_bar_image_gouwu"
            android:layout_width="27dp"
            android:layout_height="27dp"
            android:layout_above="@+id/bottom_bar_text_gouwu"
            android:layout_alignParentTop="true"
            android:layout_centerHorizontal="true"
            android:layout_marginTop="3dp"
            android:src="@drawable/icon_shoppingcarpage"/>
    </RelativeLayout>
    <RelativeLayout
        android:id="@+id/bottom_bar_weixiu_btn"
        android:layout_width="0dp"
        android:layout_height="match_parent"
        android:layout_weight="1">
        <TextView
            android:id="@+id/bottom_bar_text_weixiu"
            android:layout_width="match_parent"
            android:layout_height="wrap_content"
            android:layout_alignParentBottom="true"
            android:layout_centerHorizontal="true"
            android:layout_marginBottom="3dp"
            android:gravity="center"
            android:singleLine="true"
            android:text="维修"
            android:textColor="#666666"
            android:textSize="20sp" />
        <ImageView
            android:id="@+id/bottom_bar_image_weixiu"
            android:layout_width="27dp"
            android:layout_height="27dp"
            android:layout_above="@+id/bottom_bar_text_weixiu"
            android:layout_alignParentTop="true"
            android:layout_centerHorizontal="true"
            android:layout_marginTop="3dp"
            android:src="@drawable/icon_findpage"/>
    </RelativeLayout>
    <RelativeLayout
        android:id="@+id/bottom_bar_myinfo_btn"
        android:layout_width="0dp"
        android:layout_height="match_parent"
        android:layout_weight="1">
        <TextView
            android:id="@+id/bottom_bar_text_myinfo"
            android:layout_width="match_parent"
            android:layout_height="wrap_content"
            android:layout_alignParentBottom="true"
            android:layout_centerHorizontal="true"
            android:layout_marginBottom="3dp"
            android:gravity="center"
            android:singleLine="true"
```

```xml
                    android:text="我"
                    android:textColor="#666666"
                    android:textSize="20sp"/>
                <ImageView
                    android:id="@+id/bottom_bar_image_myinfo"
                    android:layout_width="27dp"
                    android:layout_height="27dp"
                    android:layout_above="@+id/bottom_bar_text_myinfo"
                    android:layout_alignParentTop="true"
                    android:layout_centerHorizontal="true"
                    android:layout_marginTop="3dp"
                    android:src="@drawable/icon_personalpage"/>
            </RelativeLayout>
        </LinearLayout>
    </RelativeLayout>
```

（3）底部导航功能的实现。

底部导航栏上的四个选项卡都需要设置监听器，因此 MainActivity 类需要实现 OnClickListener 接口。在 MainActivity 类中定义 setListener()方法用于为底部导航栏按钮设置监听器，并实现 OnClickListener 接口的 Onclick()方法。

底部导航栏每个选项卡都有选中和未选中两种状态，创建 setSelectedStatus()方法和 clearBottomImageState()方法来设置和清除其状态。

界面中间部分显示的视图是根据底部选项卡的选中状态进行切换的，创建 createView() 方法和 removeAllView()方法来创建视图和清除无用视图，创建 setInitStatus()方法和 selectDisplayView()方法来设置初始化的 View 界面和单击底部选项卡时对应的 View 界面。

创建 readLoginStatus()方法和 clearLoginStatus()方法用来读取和清除 SharedPreferences 中的登录状态。

如果是社区服务人员登录，则系统调用 startMyService()方法启动服务 UpdateListService，自动检测新订单，并发出处理通知。此服务会开启一个子线程，只要该社区服务人员还处于登录状态，那么每隔 10 秒查询 1 次服务器，如果发现有新的购物需求或维修需求，则会发出提醒通知服务人员接单处理。

首页相应的 MainActivity 代码如下：

```java
public class MainActivity extends AppCompatActivity implements View.OnClickListener {
    private MyInfoView mMyInfoView;
    private ShoppingView shoppingView;
    private RepairView repairView;
    private ShopallView shopallView;
    private NoLoginView noLoginView;
    private HomeView homeView;
    //repairview1 处理维修界面
    private RepairallView repairView1;
    //启动服务专用 Intent
    public static Intent serIntent=null;
    /*中间内容*/
    private FrameLayout mBodyLayout;
    /*底部导航栏*/
    private LinearLayout mBottomLayout;
    /*底部选项卡*/
```

```java
            private View mHomeBtn;
            private View mGouwuBtn;
            private View mWeixiuBtn;
            private View mMyinfoBtn;
            private TextView tv_home;
            private TextView tv_gouwu;
            private TextView tv_weixiu;
            private TextView tv_myinfo;
            private ImageView iv_home;
            private ImageView iv_gouwu;
            private ImageView iv_weixiu;
            private ImageView iv_myinfo;
            private TextView tv_back;
            private TextView tv_main_title;
            private RelativeLayout rl_title_bar;
            public static final MediaType mediaType= MediaType.parse("application/json; charset=utf-8");
            public static final String strDateFormat = "yyyy-MM-dd HH:mm:ss";
            public static final SimpleDateFormat sdf = new SimpleDateFormat(strDateFormat);
            @Override
            protected void onCreate(Bundle savedInstanceState) {
                super.onCreate(savedInstanceState);
                setContentView(R.layout.activity_main);
                if (android.os.Build.VERSION.SDK_INT > 9) {
                    StrictMode.ThreadPolicy policy = new StrictMode.ThreadPolicy.Builder().permitAll().build();
                    StrictMode.setThreadPolicy(policy);
                }
        //设置此界面为竖屏 setRequestedOrientation(ActivityInfo.SCREEN_ORIENTATION_ PORTRAIT);
                init();        //获取界面上的 UI 控件
                initBottomBar();    //获取底部导航栏上的控件
                setListener();    //底部导航栏选项卡设置监听器
                setInitStatus();    //设置界面 View 的初始化状态
                Intent intent=getIntent();
                if (intent!=null){
                    if (intent.getIntExtra("gouwu",0)==1){
                        clearBottomImageState();
                        selectDisplayView(0);
                    }else if (intent.getIntExtra("baoxiu",0)==1){
                        clearBottomImageState();
                        selectDisplayView(1);
                    }
                }
            }
            /*获取界面上的 UI 控件*/
            private void init(){
                tv_back=findViewById(R.id.tv_back);
                tv_main_title=findViewById(R.id.tv_main_title);
                rl_title_bar=findViewById(R.id.title_bar);
                rl_title_bar.setBackgroundColor(Color.parseColor("#30B4FF"));
                tv_back.setVisibility(View.GONE);
                initBodyLayout();
            }
            private void initBodyLayout(){
```

```java
    mBodyLayout=findViewById(R.id.main_body);
}
/*获取底部导航栏上的控件*/
private void initBottomBar(){
    mBottomLayout=findViewById(R.id.main_bottom_bar);
    mHomeBtn=findViewById(R.id.bottom_bar_home_btn);
    mGouwuBtn=findViewById(R.id.bottom_bar_gouwu_btn);
    mWeixiuBtn=findViewById(R.id.bottom_bar_weixiu_btn);
    mMyinfoBtn=findViewById(R.id.bottom_bar_myinfo_btn);
    tv_home=findViewById(R.id.bottom_bar_text_home);
    tv_gouwu=findViewById(R.id.bottom_bar_text_gouwu);
    tv_weixiu=findViewById(R.id.bottom_bar_text_weixiu);
    tv_myinfo=findViewById(R.id.bottom_bar_text_myinfo);
    iv_home=findViewById(R.id.bottom_bar_image_home);
    iv_gouwu=findViewById(R.id.bottom_bar_image_gouwu);
    iv_weixiu=findViewById(R.id.bottom_bar_image_weixiu);
    iv_myinfo=findViewById(R.id.bottom_bar_image_myinfo);
}
/*为底部选项卡设置监听器*/
private void setListener(){
    for (int i=0;i<mBottomLayout.getChildCount();i++){
        mBottomLayout.getChildAt(i).setOnClickListener(this);
    }
}
/*设置界面 View 的初始化状态*/
private void setInitStatus(){
    clearBottomImageState();
    int identity= AnalysisUtils.readLoginIdentity(this);
    if (identity==0){
        setSelectedStatus(0);
        createView(0);
        startMyService();   //启动服务，自动检测订单更新
        mHomeBtn.setVisibility(View.GONE);
    }else{
        setSelectedStatus(3);
        createView(3);
        startMyService();
        mHomeBtn.setVisibility(View.VISIBLE);
    }
}
/*底部导航按钮的单击事件*/
@SuppressLint("NonConstantResourceId")
@Override
public void onClick(View v) {
    switch (v.getId()) {
        //购物的单击事件
        case R.id.bottom_bar_gouwu_btn:
            clearBottomImageState();
            selectDisplayView(0);
            break;
        //维修服务的单击事件
        case R.id.bottom_bar_weixiu_btn:
            clearBottomImageState();
```

```java
                selectDisplayView(1);
                break;
            //我的单击事件
            case R.id.bottom_bar_myinfo_btn:
                clearBottomImageState();
                selectDisplayView(2);
                if (mMyInfoView != null) {
                    mMyInfoView.setLoginParams(readLoginStatus());
                }
                break;
            //主页
            case R.id.bottom_bar_home_btn:
                clearBottomImageState();
                selectDisplayView(3);
                break;
            default:
                break;
        }
    }
    /*清除底部按钮的选中状态*/
    private void clearBottomImageState(){
        tv_home.setTextColor(Color.parseColor("#666666"));
        tv_gouwu.setTextColor(Color.parseColor("#666666"));
        tv_weixiu.setTextColor(Color.parseColor("#666666"));
        tv_myinfo.setTextColor(Color.parseColor("#666666"));
        iv_home.setImageResource(R.drawable.icon_firstpage);
        iv_gouwu.setImageResource(R.drawable.icon_shoppingcarpage);
        iv_weixiu.setImageResource(R.drawable.icon_findpage);
        iv_myinfo.setImageResource(R.drawable.icon_firstpage);
        for (int i = 0; i < mBottomLayout.getChildCount(); i++) {
            mBottomLayout.getChildAt(i).setSelected(false);
        }
    }
    /*显示对应的页面*/
    private void selectDisplayView(int index){
        removeAllView();
        createView(index);
        setSelectedStatus(index);
    }
    //启动服务，自动检测订单更新
    private void startMyService(){
        if (serIntent==null){
            serIntent=new Intent(MainActivity.this, UpdateListService.class);
            startService(serIntent);
        }
    }
    /*移除不需要的视图*/
    private void removeAllView(){
        for (int i = 0; i < mBodyLayout.getChildCount(); i++) {
            mBodyLayout.getChildAt(i).setVisibility(View.GONE);
        }
    }
    /*选择视图*/
```

```java
        private void createView(int viewIndex){
//获取登录身份
            int identity= AnalysisUtils.readLoginIdentity(this);
            switch (viewIndex) {
                case 0:{
                    //购物界面
                    if (identity==1){
                        if (shoppingView==null){
                            //显示社区用户购物需求发布界面
                            shoppingView=new ShoppingView(this);
                            mBodyLayout.addView(shoppingView.getView());
                        }else{
                            shoppingView.getView();
                        }
                        shoppingView.showView();
                    }else if (identity==0){
                        //社区服务人员接单
                        if (shopallView==null){
                            //显示社区用户购物订单
                            shopallView=new ShopallView(this);
                            mBodyLayout.addView(shopallView.getView());
                        }else {
                            shopallView.getView();
                        }
                        shopallView.showView();
                    }else{
                        if (noLoginView==null){
                            noLoginView=new NoLoginView(this);
                            mBodyLayout.addView(noLoginView.getView());
                        }else {
                            noLoginView.getView();
                        }
                        noLoginView.showView();
                    }
                }
                break;
                case 1: {
                    //报修界面
                    if (identity==1){
                        //社区用户报修
                        if (repairView==null){
                            repairView=new RepairView(this);
                            mBodyLayout.addView(repairView.getView());
                        }else{
                            repairView.getView();
                        }
                        repairView.showView();
                    }else if (identity==0){
                        //社区服务人员接单
                        if (repairView1==null){
                            repairView1=new RepairallView(this);
                            mBodyLayout.addView(repairView1.getView());
                        }else{
```

```java
                            repairView1.getView();
                        }
                        repairView1.showView();
                    }else{
                        if (noLoginView==null){
                            noLoginView=new NoLoginView(this);
                            mBodyLayout.addView(noLoginView.getView());
                        }else {
                            noLoginView.getView();
                        }
                        noLoginView.showView();
                    }
                }
                break;
            case 2:
                //我的界面
                if (mMyInfoView == null) {
                    mMyInfoView = new MyInfoView(this);
                    mBodyLayout.addView(mMyInfoView.getView());
                } else {
                    mMyInfoView.getView();
                }
                mMyInfoView.showView();
                break;
            case 3:
                if (homeView==null){
                    homeView=new HomeView(this);
                    mBodyLayout.addView(homeView.getView());
                }else {
                    homeView.getView();
                }
                homeView.showView();
                break;
        }
    }
    /*设置底部选项卡的选中状态*/
    private void setSelectedStatus(int index){
        int identity= AnalysisUtils.readLoginIdentity(this);
        switch (index) {
            case 0:
                mGouwuBtn.setSelected(true); iv_gouwu.setImageResource (R.drawable.icon_shoppingcarclicked);
                tv_gouwu.setTextColor(Color.parseColor("#0097F7"));
                rl_title_bar.setVisibility(View.VISIBLE);
                if (identity==0){
                    tv_main_title.setText("购物处理(社区服务人员)");
                }else if (identity==1){
                    tv_main_title.setText("购物发布");
                }else {
                    tv_main_title.setText("未登录");
                }
                break;
            case 1:
```

```java
                    mWeixiuBtn.setSelected(true);
                    iv_weixiu.setImageResource(R.drawable.icon_findpageclicked);
                    tv_weixiu.setTextColor(Color.parseColor("#0097F7"));
                    rl_title_bar.setVisibility(View.VISIBLE);
                    if (identity==0){
                        tv_main_title.setText("报修处理(社区服务人员)");
                    }else if (identity==1){
                        tv_main_title.setText("报修提交");
                    }else {
                        tv_main_title.setText("未登录");
                    }
                    break;
            case 2:
                    mMyinfoBtn.setSelected(true);
                    iv_myinfo.setImageResource(R.drawable.icon_firstpageclicked);
                    tv_myinfo.setTextColor(Color.parseColor("#0097F7"));
                    rl_title_bar.setVisibility(View.GONE);
                    break;
            case 3:
                    //主页
                    mHomeBtn.setSelected(true);
                    iv_home.setImageResource(R.drawable.icon_findpageclicked);
                    tv_home.setTextColor(Color.parseColor("#0097F7"));
                    rl_title_bar.setVisibility(View.GONE);
            }
    }
    //清空视图
    public void clearView(){
        homeView=null;
        mMyInfoView=null;
        shoppingView=null;
        repairView=null;
        shopallView=null;
        repairView1=null;
        noLoginView=null;
    }
    @Override
    protected void onActivityResult(int requestCode, int resultCode, Intent data) {
        super.onActivityResult(requestCode, resultCode, data);
        if(data!=null){
            //从设置界面或登录界面传递过来的登录状态
            boolean isLogin=data.getBooleanExtra("isLogin",false);
            if(isLogin){    //登录成功时显示的界面
                clearBottomImageState();
                clearView();
                int identity= AnalysisUtils.readLoginIdentity(this);
                if (identity==0){
                    mHomeBtn.setVisibility(View.GONE);
                    selectDisplayView(0);
                }else if (identity==1){
                    mHomeBtn.setVisibility(View.VISIBLE);
                    selectDisplayView(3);
                }
```

```java
            }else{
                mHomeBtn.setVisibility(View.VISIBLE);
            }
//登录成功或退出登录时根据 isLogin 设置我的界面
            if (mMyInfoView != null) {                              mMyInfoView.setLoginParams(isLogin);
            }
        }
    }
    protected long exitTime;    //记录第一次单击时的时间
    @Override
    public boolean onKeyDown(int keyCode, KeyEvent event){
        if (keyCode == KeyEvent.KEYCODE_BACK
                && event.getAction() == KeyEvent.ACTION_DOWN) {
            if ((System.currentTimeMillis() - exitTime) > 2000) {
                Toast.makeText(MainActivity.this, "再按一次退出服务社区",
                        Toast.LENGTH_SHORT).show();
                exitTime = System.currentTimeMillis();
            } else {
                MainActivity.this.finish();
                if (readLoginStatus()) {
//如果退出此应用时是登录状态,则需要清除登录状态
//同时需清除登录时的用户名
                    clearLoginStatus();
                    if (serIntent!=null){
                        stopService(serIntent);
                        serIntent=null;
                    }
                }
                clearView();
                System.exit(0);
                finishActivity(0);
            }
            return true;
        }
        return super.onKeyDown(keyCode, event);
    }
    /**
     * 获取 SharedPreferences 中的登录状态
     */
    private boolean readLoginStatus() {
        SharedPreferences sp = getSharedPreferences("loginInfo",
                Context.MODE_PRIVATE);
        return sp.getBoolean("isLogin", false);
    }
    /**
     * 清除 SharedPreferences 中的登录状态
     */
    private void clearLoginStatus() {
        System.out.println("clear");
        SharedPreferences sp = getSharedPreferences("loginInfo", Context.MODE_PRIVATE);
        SharedPreferences.Editor editor = sp.edit();    //获取编辑器
```

```
            editor.clear();
            editor.commit();    //提交修改
        }
    }
```

2. 登录界面

登录界面（见图 10.2），单选按钮用于选择登录身份，系统可根据用户选择的身份跳转到对应的用户界面。单击"登录"按钮后系统验证手机号和密码，登录成功后跳转到首页界面。登录界面上提供了一个"立即注册"的文本区域，单击后根据当前选择的用户身份跳转到对应注册界面。

登录界面的布局比较简单，不再详细介绍，读者可以参考本书提供的相关代码。

登录时首先验证用户输入的信息是否完整，完整性验证通过后，使用 OkHttp 构造 POST 请求连接服务器进行账号密码的正确性验证，使用 MultipartBody.Builder 类存储请求中的参数，服务器端通过@RequestParam 注解接收参数，通过调用服务层方法获取数据库数据，验证后返回结果。客户端接收结果并做出对应的操作，验证成功后返回用户数据，将此数据存入本地文件。

登录 LoginActivity 对应的代码如下：

```
public class LoginActivity extends AppCompatActivity {
    private TextView tv_main_title;
    private TextView tv_back,tv_register,tv_find_psw;
    private Button btn_login;
    private String phone,psw;
    private EditText et_user_phone,et_psw;
    private RadioGroup identity;
    private int user_identity=1;
    @Override
    protected void onCreate(Bundle savedInstanceState) {
        super.onCreate(savedInstanceState);
        setContentView(R.layout.activity_login);
        //设置此界面为竖屏
        setRequestedOrientation(ActivityInfo.SCREEN_ORIENTATION_PORTRAIT);
        init();
    }
    /*获取界面控件*/
    private void init(){
        tv_main_title=findViewById(R.id.tv_main_title);
        tv_main_title.setText("登录");
        tv_back=findViewById(R.id.tv_back);
        tv_register=findViewById(R.id.tv_register);
        tv_find_psw=findViewById(R.id.tv_find_psw);
        btn_login=findViewById(R.id.btn_login);
        et_user_phone=findViewById(R.id.et_user_phone);
        et_psw=findViewById(R.id.et_psw);
        identity=findViewById(R.id.identity);
        identity.setOnCheckedChangeListener(new      RadioGroup.OnCheckedChangeListener() {
            @Override
            public void onCheckedChanged(RadioGroup group, int checkedId) {
                RadioButton radioButton=findViewById(checkedId);
                if (radioButton==null){
```

```java
                    user_identity=-1;
                }else {
                    String result=radioButton.getText().toString();
                    if (result.equals("服务人员")){
                        user_identity=0;
                    }else if (result.equals("住户")){
                        user_identity=1;
                    }
                }
            }
    });
    //返回按钮的单击事件
    tv_back.setOnClickListener(new View.OnClickListener() {
        @Override
        public void onClick(View v) {
            LoginActivity.this.finish();
        }
    });
    //立即注册控件的单击事件
    tv_register.setOnClickListener(new View.OnClickListener() {
        @Override
        public void onClick(View v) {
            if (user_identity==1){
                //社区住户注册
                Intent intent=new Intent(LoginActivity.this,RegisterActivity.class);
                startActivityForResult(intent,1);
            }else {
                //社区服务人员注册
                Intent intent=new Intent(LoginActivity.this,RegisterActivity2.class);
                startActivityForResult(intent,0);
            }
        }
    });
    //找回密码控件单击事件
    tv_find_psw.setOnClickListener(new View.OnClickListener() {
        @Override
        public void onClick(View v) {
            //跳转到找回密码界面
        }
    });
    //登录按钮的单击事件
    btn_login.setOnClickListener(new View.OnClickListener() {
        @Override
        public void onClick(View v) {
            phone=et_user_phone.getText().toString().trim();
            psw=et_psw.getText().toString().trim();
            //完整性验证
            if (TextUtils.isEmpty(phone)){
                Toast.makeText(LoginActivity.this,
                    "请输入手机号",Toast.LENGTH_SHORT).show();
            }else if (TextUtils.isEmpty(psw)){
                Toast.makeText(LoginActivity.this,
                    "请输入密码",Toast.LENGTH_SHORT).show();
```

```java
        } else if(user_identity==-1){
            Toast.makeText(LoginActivity.this,
                "请选择身份",Toast.LENGTH_SHORT).show();
        }else{
            String ipaddress=getResources().getString(R.string.ipaddress);
            String url;
            //社区住户登录
            if (user_identity==1) {
                url = ipaddress + "findUserByPhone";
            }else {
                //社区服务人员登录
                url = ipaddress + "findServiceUserByPhone";
            }
            //OkHttp 请求服务器
            MultipartBody.Builder urlBuilder = new     MultipartBody
                    .Builder() .setType(MultipartBody.FORM);
            urlBuilder.addFormDataPart("phone", phone);
            urlBuilder.addFormDataPart("pwd", psw);
            Request request = new Request.Builder()
                    .url(url)
                    .post(urlBuilder.build())
                    .build();
            Call call = client.newCall(request);
            try {
                Response response = call.execute();
                if (response.isSuccessful()) {
                    assert response.body() != null;
                    String message = response.body().string();
                    if (user_identity==1){
                        User user = JSON.parseObject(message, User.class);
                        if (user != null) {
                            //登录成功，保存用户数据
                SharedPreferences sp=getSharedPreferences("loginInfo",
                    MODE_PRIVATE);
                SharedPreferences.Editor editor=sp.edit();
                    //获取编辑器
                editor.putBoolean("isLogin",true);
                editor.putInt("identity",1);
                editor.putInt("id", user.getId());
                //存入登录时的社区用户
                editor.putString("username",user.getUsername());
                editor.putString("nickname",user.getNickname());
                editor.putString("pwd",user.getPwd());
                editor.putString("phone",user.getPhone());
                editor.putString("address",user.getAddress());
                editor.putString("number",user.getNumber());
                editor.putString("pic",user.getPic());
                editor.apply();
                Intent data = new Intent();
                data.putExtra("isLogin", true);
                data.putExtra("identity",1);
                setResult(RESULT_OK, data);
                LoginActivity.this.finish();
```

```java
            } else {
                Toast.makeText(LoginActivity.this, "用户名或密码错误",
                    Toast.LENGTH_SHORT).show();
                }
            }else{
                ServiceUser serviceUser = JSON.parseObject(message,
            ServiceUser.class);
                if (serviceUser != null) {
SharedPreferences sp=getSharedPreferences("loginInfo",MODE_PRIVATE);
SharedPreferences.Editor editor=sp.edit();       //获取编辑
editor.putBoolean("isLogin",true);
editor.putInt("identity",0);
editor.putInt("id", serviceUser.getId());                //存入登录时的社区服务人员
editor.putString("name", serviceUser.getName());
editor.putString("nickname", serviceUser.getNickname());
editor.putString("pwd", serviceUser.getPwd());
editor.putString("phone",serviceUser.getPhone());
editor.putString("number",serviceUser.getNumber());
editor.putString("pic",serviceUser.getPic());
editor.apply();
Intent data = new Intent();
data.putExtra("isLogin", true);
data.putExtra("identity",0);
setResult(RESULT_OK, data);
LoginActivity.this.finish();
} else {
Toast.makeText(LoginActivity.this, "用户名或密码错误", Toast.LENGTH_SHORT).show();
 }
 }
 }
 } catch (IOException e) {
     e.printStackTrace();
}
 }
 }
});
        }
        @Override
        protected void onActivityResult(int requestCode,int resultCode,Intent data){
            super.onActivityResult(requestCode,resultCode,data);
            if (data!=null){
                String phone=data.getStringExtra("phone");
                if (!TextUtils.isEmpty(phone)){
                    et_user_phone.setText(phone);
                    //设置光标位置
                    et_user_phone.setSelection(phone.length());
                    et_psw.setText("");
                }
                if (requestCode==0){
                    RadioButton r1=findViewById(R.id.radio1);
                    r1.setChecked(true);
                }
            }
        }
}
```

3. 购物和维修需求发布界面

社区住户购物界面和报修界面（见图 10.5 和图 10.6）的结构相同，上部是标题栏，下部是导航栏，中间部分为一个自定义的 View，根据社区住户选择的导航栏选项卡加载不同的 View。

单击底部导航栏的"购物"或"维修"选项卡时，系统首先移除主体部分的视图内容，同时为方便再次调用，加快程序的响应速度，仅将其设为不可见而不是释放资源，移除现有视图后，加载相应的视图。

以下是购物界面对应的 ShoppingView 的代码：

```java
public class ShoppingView {
    private EditText et_coname, et_num, et_remark;
    private Button bt_submit;
    private View mCurrentView;
    private Activity mContext;
    private LayoutInflater mInflater;
    private String coname, num, remark;
    public ShoppingView(Activity context) {
        mContext = context;
        //为之后将 Layout 转化为 View 时用
        mInflater = LayoutInflater.from(mContext);
    }
    private void createView() {
        initView();
    }
    @SuppressLint("InflateParams")
    public void initView() {
        mCurrentView = mInflater.inflate(R.layout.shopping, null);
        et_coname = mCurrentView.findViewById(R.id.et_coname);
        et_num = mCurrentView.findViewById(R.id.et_num);
        et_remark = mCurrentView.findViewById(R.id.et_remark);
        bt_submit = mCurrentView.findViewById(R.id.bt_submit);
        bt_submit.setOnClickListener(v -> {
            getEditString();
            if (TextUtils.isEmpty(coname)) {
                Toast.makeText(mContext, "请输入商品名", Toast.LENGTH_SHORT).show();
            } else if (TextUtils.isEmpty(num)) {
                Toast.makeText(mContext, "请输入商品数量", Toast.LENGTH_SHORT).show();
            } else {
                AlertDialog.Builder normalDialog =
                        new AlertDialog.Builder(mContext);
                normalDialog.setTitle("确认购买");
                normalDialog.setMessage("您确定要购买" + coname + " x " + num + "吗？\n 我们将会根据您的个人信息安排服务");
                normalDialog.setPositiveButton("确定",
                        (dialog, which) -> {
                    String ipaddress = mContext.getResources().getString(R.string.ipaddress);
                    String url = ipaddress + "addPurchase";
                    OkHttpClient client = new OkHttpClient();
                    Purchase purchase = new Purchase();
                    getInfo(purchase);
                    String comark = coname + " " + remark;
```

```java
                    purchase.setConame(comark.trim());
                    purchase.setNum(Integer.valueOf(num));
                    String purchaseinfo = JSON.toJSONString(purchase);
                    Request request = new Request.Builder().url(url)
                                .post(RequestBody.create(mediaType, purchaseinfo))
                                .build();
                    Call call = client.newCall(request);
                    try {
                        Response response = call.execute();
                        if (response.isSuccessful()) {
                            assert response.body() != null;
                            String message = response.body().string();
                            if (message.equals("true")) {
                                Toast.makeText(mContext, "提交成功", Toast.LENGTH_SHORT).show();
                            et_coname.setText("");
                            et_num.setText("");
                            et_remark.setText("");
                            Intent intent=new Intent(mContext, Historypill.class);
                            intent.putExtra("re",1);
                                mContext.startActivity(intent);
        } else {
                Toast.makeText(mContext, "提交失败", Toast.LENGTH_SHORT).show();
            }
        }
        } catch (IOException e) {
                e.printStackTrace();
    }
});
normalDialog.setNegativeButton("关闭",
                            (dialog, which) -> dialog.dismiss());
normalDialog.show();
}
});
}
    public View getView() {
        if (mCurrentView == null) {
            createView();
        }
        return mCurrentView;
    }

    public void showView() {
        if (mCurrentView == null) {
            createView();
        }
        mCurrentView.setVisibility(View.VISIBLE);
    }

    private void getEditString() {
        coname = et_coname.getText().toString().trim();
        num = et_num.getText().toString().trim();
        remark = et_remark.getText().toString().trim();
    }
```

```java
        private void getInfo(Purchase purchase) {
            SharedPreferences sp = mContext.getSharedPreferences("loginInfo", Context.MODE_PRIVATE);
            purchase.setUserid(sp.getInt("id", 0));
            purchase.setUsername(sp.getString("nickname", ""));
            purchase.setUserphone(sp.getString("phone", ""));
        }
    }
```

4. 查看历史购物记录

用户查看历史购物记录界面（见图 10.8），界面由顶部订单状态选择栏、中部展示列表和底部返回按钮组成。用户可以选择查看所有订单或对订单状态进行筛选，根据身份的不同，社区住户可选择未接单、已接单、服务中、已完成和已评价状态。

通过 getPurchaseByUserId()方法获取当前登录用户的所有购物记录，并按订单状态分别保存在相应的 List 集合中，然后以列表形式显示在界面上。

getPurchaseByUserId()方法主要代码如下：

```java
            private void getPurchaseByUserId(int userid) {
                String ipaddress = getResources().getString(R.string.ipaddress);
                String url = ipaddress + "/findPurchaseByUserId";
                OkHttpClient client = new OkHttpClient();
                MultipartBody.Builder urlBuilder = new MultipartBody.Builder()
                        .setType(MultipartBody.FORM);
                urlBuilder.addFormDataPart("userid", String.valueOf(userid));
                Request request = new Request.Builder()
                        .url(url)
                        .post(urlBuilder.build())
                        .build();
                Call call = client.newCall(request);
                try {
                    Response response = call.execute();
                    if (response.isSuccessful()) {
                        assert response.body() != null;
                        String message = response.body().string();
                        listall = JSONArray.parseArray(message, Purchase.class);
                        listrece= new ArrayList<>();
                        liststart= new ArrayList<>();
                        listover= new ArrayList<>();
                        listeva= new ArrayList<>();
                        listend= new ArrayList<>();
                        for (Purchase p:listall){
                            switch (p.getStatus()) {
                                case "未接单":
                                    listrece.add(p);
                                    break;
                                case "已接单":
                                    liststart.add(p);
                                    break;
                                case "服务中":
                                    listover.add(p);
                                    break;
                                case "已完成":
```

```java
                                    listeva.add(p);
                                    break;
                                case "已评价":
                                    listend.add(p);
                                    break;
                                default:
                                    System.out.println("statuserror");
                                    break;
                            }
                        }
                    }
            } catch (IOException e) {
                e.printStackTrace();
            }
        }
        //设置 ListView
        private void setListView(List<Purchase> list){
            List<HashMap<String,Object>>data=new ArrayList<HashMap<String,Object>>();
            for (Purchase p:list){
                HashMap<String, Object>item=new HashMap<String, Object>();
                item.put("coname",p.getConame());
                item.put("num",p.getNum());
                if (p.getStatus().equals("未接单")){
                    item.put("name","未接单");
                    item.put("phone","未接单");
                }else{
                    item.put("name",p.getServicename());
                    item.put("phone",getPhoneByServiceUserId(p.getServiceid()));
                }
                item.put("status",p.getStatus());
                String s=p.getStatus();
                switch (s) {
                    case "未接单":
                        item.put("time1", "发布时间：");
                        item.put("time2", sdf.format(p.getCreatetime()));
                        break;
                    case "已接单":
                        item.put("time1", "接单时间：");
                        item.put("time2", sdf.format(p.getRecetime()));
                        break;
                    case "服务中":
                        item.put("time1", "开始时间：");
                        item.put("time2", sdf.format(p.getStarttime()));
                        break;
                    case "已完成":
                        item.put("time1", "完成时间：");
                        item.put("time2", sdf.format(p.getOvertime()));
                        break;
                    case "已评价":
                        item.put("time1", "评价时间：");
                        item.put("time2", sdf.format(p.getEvatime()));
                        break;
                }
```

```
            data.add(item);
        }
        SimpleAdapter simpleAdapter=new SimpleAdapter(this,data,R.layout.shoplist,
            new String[]{"coname","num","name","phone","status","time1","time2"},
            new int[]{R.id.tw_coname,R.id.tw_num,R.id.tw_serv_name,
                    R.id.tw_phonne,R.id.tw_status,R.id.tw_time1,R.id.tw_time2});
        lv_shopping.setAdapter(simpleAdapter);

    private String getPhoneByServiceUserId(int serviceuserid){
        String ipaddress = getResources().getString(R.string.ipaddress);
        String url = ipaddress + "findServiceUserById";
        OkHttpClient client = new OkHttpClient();
        MultipartBody.Builder urlBuilder = new MultipartBody.Builder()
                .setType(MultipartBody.FORM);
        urlBuilder.addFormDataPart("id", String.valueOf(serviceuserid));
        Request request = new Request.Builder()
                .url(url)
                .post(urlBuilder.build())
                .build();
        Call call = client.newCall(request);
        try {
            Response response = call.execute();
            if (response.isSuccessful()) {
                assert response.body() != null;
                String message = response.body().string();
                ServiceUser user=JSON.parseObject(message,ServiceUser.class);
                return user.getPhone();
            }else{
                return null;
            }
        } catch (IOException e) {
            e.printStackTrace();
            return null;
        }
    }
```

5．服务人员接单界面

服务人员接单界面（见图 10.11），通过界面服务人员可以看到购物需求的商品名、数量等基本信息，在社区服务人员接单后双方都可以在历史购物订单界面查看到订单信息，并显示彼此的手机号，社区服务人员接单后 30 分钟内必须开始服务，否则会收到系统的提示通知。

主要代码如下：

```
//获取所有状态为未接单的商品
private void getStatusPur(){
    String ipaddress = mContext.getResources().getString(R.string.ipaddress);
    String url = ipaddress + "findPurchaseByStatus";
    OkHttpClient client = new OkHttpClient();
    MultipartBody.Builder urlBuilder = new MultipartBody.Builder()
            .setType(MultipartBody.FORM);
    urlBuilder.addFormDataPart("status", "未接单");
    Request request = new Request.Builder()
```

```java
                                .url(url)
                                .post(urlBuilder.build())
                                .build();
                    Call call = client.newCall(request);
                    try {
                        Response response = call.execute();
                        if (response.isSuccessful()) {
                            assert response.body() != null;
                            String message = response.body().string();
                            list= JSONArray.parseArray(message,Purchase.class);
                        }
                    } catch (IOException e) {
                        e.printStackTrace();
                    }
                }

        //设置 ListView
        private void setListView(){
            List<HashMap<String,Object>> data=new ArrayList<HashMap<String,Object>>();
            for (Purchase p:list){
                HashMap<String, Object>item=new HashMap<String, Object>();
                item.put("coname",p.getConame());
                item.put("num",p.getNum());
                item.put("address",getAddressById(p.getUserid()));
                item.put("status",p.getStatus());
                item.put("time", sdf.format(p.getCreatetime()));
                data.add(item);
            }
            SimpleAdapter simpleAdapter=new SimpleAdapter(mContext,data,R.layout.shopalllist,
                new String[]{"coname","num","address","status","time"},
                new int[]{R.id.tw_coname,R.id.tw_num,R.id.tw_address,        R.id.tw_status,R.id.tw_time});
            lv_shopall.setAdapter(simpleAdapter);
lv_shopall.setOnItemClickListener(new AdapterView.OnItemClickListener() {

                @Override
                public void onItemClick(AdapterView<?> parent, View view, int position, long id) {
                    Purchase purchase=list.get(position);
                    AlertDialog.Builder normalDialog =
                            new AlertDialog.Builder(mContext);
                    normalDialog.setTitle("接单");
                    normalDialog.setMessage("是否接收此业务，接收后需在 30 分钟内联系社区住户进行服务");
                    normalDialog.setPositiveButton("确定",
                            (dialog, which) -> {

purchase.setServiceid(AnalysisUtils.getLoginId(mContext));
purchase.setServicename(AnalysisUtils.getLoginName(mContext));
purchase.setStatus("已接单");
String ipaddress=mContext.getResources().getString(R.string.ipaddress);
String url=ipaddress+"updatePurchaseStatus";
OkHttpClient client=new OkHttpClient();
String purinfo= JSON.toJSONString(purchase);
Request request=new Request.Builder()
```

```java
                                        .url(url)
                                        .post(RequestBody.create(mediaType,purinfo))
                                        .build();
        Call call=client.newCall(request);
         try {
                Response response=call.execute();
                if (response.isSuccessful()){
                    assert response.body() != null;
                    String message=response.body().string();
                    if (message.equals("true")){
                       //30 分钟后未处理发出通知
                        getChildThread(purchase.getId());
                        Toast.makeText(mContext,"接单成功",Toast.LENGTH_SHORT).show();
                    Intent intent=new Intent(mContext, PurchaseServer.class);
                    intent.putExtra("re",1);
                  mContext.startActivity(intent);
    }else {
         Toast.makeText(mContext,"接单失败",Toast.LENGTH_SHORT).show();
   }
     getStatusPur();
     setListView();
   }
 } catch (IOException e) {
         e.printStackTrace();
    }
});
normalDialog.setNegativeButton("取消",
              (dialog, which) -> dialog.dismiss());
normalDialog.show();
  }
});
 }
  //根据用户 ID 获取用户地址
     private String getAddressById(int userid){
            String ipaddress = mContext.getResources().getString(R.string.ipaddress);
            String url = ipaddress + "findUserById";
            OkHttpClient client = new OkHttpClient();
            MultipartBody.Builder urlBuilder = new MultipartBody.Builder()
                    .setType(MultipartBody.FORM);
            urlBuilder.addFormDataPart("id", String.valueOf(userid));
            Request request = new Request.Builder()
                    .url(url)
                    .post(urlBuilder.build())
                    .build();
            Call call = client.newCall(request);
            try {
                Response response = call.execute();
                if (response.isSuccessful()) {
                    assert response.body() != null;
                    String message = response.body().string();
                    User user=JSON.parseObject(message,User.class);
                    return user.getAddress();
                }else{
                    return null;
                }
```

```
        } catch (IOException e) {
            e.printStackTrace();
            return null;
        }
    }
```

6. 新订单提醒

社区住户发布购物或维修需求后,系统会向社区服务人员发出通知,提醒社区服务人员尽快接单服务(见图 10.13)。该功能通过 Notification 状态栏通知实现。首先创建一个子线程,实时监测社区住户发布的新订单,进而调用 Notification 状态栏通知向社区服务人员发送通知。

核心代码:

```
        @Override
        public void run() {
            try {
                getSum(0);
                System.out.println("re:"+reSum);
                System.out.println("pur:"+purSum);
                while(MainActivity.serIntent!=null){
                    Thread.sleep(10000);
                    getSum(1);
                    System.out.println("rex:"+reSumx);
                    System.out.println("purx:"+purSumx);
                    if (reSumx>reSum){
                        //发出维修通知
                        putNot(2,reSumx-reSum);
                        reSum=reSumx;
                    }
                    if (purSumx>purSum){
                        //发出购物通知
                        putNot(1,purSumx-purSum);
                        purSum=purSumx;
                    }
                }
            } catch (InterruptedException e) {
                e.printStackTrace();
            }
        }
    }.start();
    return super.onStartCommand(intent, flags, startId);
}
    //zl 是指发送的通知是维修还是购物,sum 是指变更的数量
private void putNot(int zl,int sum){
    String contextTitle="新订单";
    String contextText;
    Intent intent=new Intent(UpdateListService.this, MainActivity.class);
    int ID;
    if (zl==1){
        contextText="新发布了"+ sum +"条购物需求";
        ID=300;
        intent.putExtra("gouwu",1);
    }else {
        contextText="新发布了"+ sum +"条维修需求";
```

```java
                ID=400;
                intent.putExtra("baoxiu",1);
        }
            NotificationManager    manager=(NotificationManager)    getSystemService(NOTIFICATION_SERVICE);
            Notification notification;
            if (Build.VERSION.SDK_INT>=Build.VERSION_CODES.O){
                String id="newDing";
                String description="is-newDing";
                int importance=NotificationManager.IMPORTANCE_LOW;
                NotificationChannel channel=new NotificationChannel(id,description,importance);
                manager.createNotificationChannel(channel);
                PendingIntent pendingIntent = PendingIntent.getActivity(UpdateListService.this, ID, intent, PendingIntent.FLAG_UPDATE_CURRENT);
                notification=new Notification.Builder(UpdateListService.this,id)
                        .setCategory(Notification.CATEGORY_MESSAGE)
                        .setSmallIcon(R.drawable.ic_launcher_background)
                        .setContentTitle(contextTitle)
                        .setContentText(contextText)
                        .setContentIntent(pendingIntent)
                        .setAutoCancel(true)
                        .build();
                manager.notify(ID,notification);
            }else{
                notification=new Notification.Builder(UpdateListService.this)
                        .setSmallIcon(R.drawable.ic_launcher_background)
                        .setContentTitle(contextTitle)
                        .setContentText(contextText)
                        .setAutoCancel(true)
                        .build();
                manager.notify(ID,notification);
            }
        }
}
```

7．服务超时提醒

社区服务人员接单超时提醒功能的通知栏设置方法与发布新需求通知栏的方法一致，都使用了 Notification 状态栏通知完成对社区服务人员的提醒。接单超时提醒则是在社区服务人员接单之后创建子线程，并使用定时器完成对社区服务人员服务状态的定时监视功能，在一定时间内如果依旧没有完成则发送通知栏通知，完成对社区服务人员的超时提醒。

核心代码如下：

```java
//开启子线程
private void getChildThread(Integer id){
    new Thread(new Runnable() {
        @Override
        public void run() {
            try {
                System.out.println("进入");
                Thread.sleep(10000);
                System.out.println("计时结束");
                String ipaddress= mContext.getResources().getString(R.string.ipaddress);
                String url=ipaddress+"findPurchaseById";
                OkHttpClient client = new OkHttpClient();
                MultipartBody.Builder urlBuilder = new MultipartBody.Builder()
```

```java
                            .setType(MultipartBody.FORM);
                    urlBuilder.addFormDataPart("id", String.valueOf(id));
                    Request request = new Request.Builder()
                            .url(url)
                            .post(urlBuilder.build())
                            .build();
                    Call call = client.newCall(request);
                    Response response = call.execute();
                    if (response.isSuccessful()) {
                        assert response.body() != null;
                        String message = response.body().string();
                        Purchase purchase=JSON.parseObject(message,Purchase.class);
                        if (purchase.getStatus().equals("已接单")){
                            sendNotice();   //发出通知}
                    }
                } catch (InterruptedException | IOException e) {
                    e.printStackTrace(); } }
        }).start();}
//发通知
    private void sendNotice(){
        String contextTitle="提醒";
        String contextText="请尽快服务";
        int ID=100;
        NotificationManager manager=(NotificationManager) mContext.getSystemService(NOTIFICATION_SERVICE);
        Notification notification;
        if (Build.VERSION.SDK_INT>=Build.VERSION_CODES.O){
            String id="purchase";
            String description="my-purchase";
            int importance=NotificationManager.IMPORTANCE_LOW;
            NotificationChannel channel=new NotificationChannel(id,description,importance);
            manager.createNotificationChannel(channel);
            Intent intent=new Intent(mContext, PurchaseServer.class);
            PendingIntent pendingIntent = PendingIntent.getActivity(mContext, ID, intent, PendingIntent.FLAG_UPDATE_CURRENT);
            notification=new Notification.Builder(mContext,id)
                    .setCategory(Notification.CATEGORY_MESSAGE)    //设置通知所属类别
                    .setSmallIcon(R.drawable.ic_launcher_background)
                    .setContentTitle(contextTitle)
                    .setContentText(contextText)
                    .setContentIntent(pendingIntent)
                    .setAutoCancel(true)
                    .build();
            manager.notify(ID,notification);
        }else{
            notification=new Notification.Builder(mContext)
                    .setSmallIcon(R.drawable.ic_launcher_background)
                    .setContentTitle(contextTitle)
                    .setContentText(contextText)
                    .setAutoCancel(true)
                    .build();
            manager.notify(ID,notification);
        }
    }
}
```